NEEDS ASSESSMENT

NEEDS ASSESSMENT

PHASE II
Collecting Data

James W. Altschuld
The Ohio State University

Series Editor: James W. Altschuld

NEEDS ASSESSMENT KIT 3

Los Angeles | London | New Delhi
Singapore | Washington DC

For information:

SAGE Publications, Inc.
2455 Teller Road
Thousand Oaks,
    California 91320
E-mail: order@sagepub.com

SAGE Publications India Pvt. Ltd.
B 1/I 1 Mohan Cooperative
    Industrial Area
Mathura Road, New Delhi 110 044
India

SAGE Publications Ltd.
1 Oliver's Yard
55 City Road
London EC1Y 1SP
United Kingdom

SAGE Publications Asia-Pacific
    Pte. Ltd.
33 Pekin Street #02-01
Far East Square
Singapore 048763

Printed in the United States of America.

Library of Congress Cataloging-in-Publication Data

Altschuld, James W.
Needs assessment phase II: collecting data/James William Altschuld.
    p. cm.
Includes bibliographical references and index.
ISBN 978-1-4129-7513-1 (pbk.)
    1. Strategic planning. 2. Needs assessment. I. Title.

HD30.28.A3885 2010
658.4'012—dc22   2009015173

This book is printed on acid-free paper.

09   10   11   12   13   10   9   8   7   6   5   4   3   2   1

Acquisitions Editor:	Vicki Knight
Associate Editor:	Sean Connelly
Editorial Assistant:	Lauren Habib
Production Editor:	Brittany Bauhaus
Copy Editor:	Melinda Masson
Typesetter:	C&M Digitals (P) Ltd.
Proofreader:	Victoria Reed-Castro
Indexer:	Diggs Publication Services, Inc.
Cover Designer:	Candice Harman
Marketing Manager:	Stephanie Adams

Brief Contents

Detailed Contents

Preface

Throughout the KIT, stress has been placed on a three-phase needs assessment process. From that perspective, Phase II (Assessment—Collecting Data) may not be necessary based upon what has been learned before. Don't automatically think that seeking new data in a second phase is the way to go. It is costly and time consuming and may not yield much additional information. The admonition is that you only enter into such work if it will be worthwhile.

If more data are needed and a decision is made to go forward, there are many methodological options depending on the specific nature of the needs assessment, the content area or focus, and knowledge of what methods and information might be most persuasive and attended to in the setting. It would not be useful to try to cover all methods in this text, so a selected set of commonly employed ones (quantitative and qualitative) is profiled in some detail. The two types of approaches are important as needs may be complicated and best understood from the perspectives of numbers, perceptions, and value positions. The interplay of findings almost always leads to deeper levels of meaning.

That being said, more methods require greater financial, time, and skill resources. Do the best you can in your assessment and consider adopting and/or adapting methods to fit what can be done within the local context and the parameters available to the needs assessment committee. It's not a perfect world, and flexibility in conducting assessments is essential.

Also, you are strongly encouraged to search (even if only skimming) the literature for what others have done with related needs in similar organizations. What methods did they choose, what methodological twists have they implemented, what problems did they run into, how successful were their applications, what recommendations do they offer, and how could what they have done be applied to your needs assessment? There is much writing about methods that should be helpful; take advantage of it!

As indicated in its title, this is Book 3 in the needs assessment KIT. The others are:

Book 1: *Needs Assessment: An Overview*

Book 2: *Phase I: Getting Started*

Book 4: *Analysis and Prioritization*

Book 5: *Phase III: Taking Action for Change*

Reference, when appropriate in this text, will be made to other books in the KIT. If you need more ideas on how to implement an assessment, you are encouraged to consult them.

Acknowledgments

Over the years, my work has been vastly improved by graduate students in my classes, doctoral candidates (now graduates) whom I was fortunate to guide through their studies and research, and other colleagues involved in numerous projects. Their insights and skeptical questioning often led to improved instrumentation and the collection of higher-quality data. Those contributions should be apparent in the numerous citations in the book. Barbara Heinlein as usual through her word-processing expertise, and keen eye made for a vastly improved text. In this regard, the help of Traci Lepicki and Stephanie Tischendorf is also appreciated. Thanks to all of these individuals for their kind and thoughtful assistance.

One thing that I have never publicly done is to state appreciation for my deceased parents, Harry and Josephine Wodicka Altschuld, and their devotion to their children, my brother (Glenn), my sisters (Rae and June), and me. They came to the United States with little education to make a better life for themselves and to experience a world free of the persecution they had seen as members of a minority group in Europe. Everything that I and my siblings have accomplished stems from what they were able to achieve. With the humblest of love and respect, thank you from the bottom of my heart.

The author and SAGE gratefully acknowledge the contributions of the following reviewers:

Evan Abbott, *Regis University*

Ellen Darden, *Concord University*

Doug Leigh, *Pepperdine University*

Wendy Lewandowski, *Kent State University*

Steven E. Meier, *University of Idaho*

Jennifer Piver-Renna, *Johns Hopkins Bloomberg School of Public Health*

Kui-Hee Song, *California State University–Chico*

About the Author

James W. Altschuld, PhD, received his bachelor's and master's degrees in chemistry from Case Western Reserve University and The Ohio State University (OSU), respectively. His doctorate is from the latter institution with an emphasis on educational research and development and sociological methods. He is now professor emeritus in the College of Education and Human Ecology at OSU after 27 years of teaching research techniques and program evaluation. In evaluation, he developed and taught a sequence of courses on theory, needs assessment, and design. He has coauthored three previous books (two on needs assessment and the other on the evaluation of science and technology education), has written many chapters on needs assessment as well as others on evaluation research and issues, and has an extensive list of publications, almost all in the field of evaluation. He has given presentations and done work in five countries outside of the United States. In his career he has been the recipient of local, state, and national honors including the Alva and Gunnar Myrdal Practice Award from the American Evaluation Association for contributions to evaluation.

1

Getting Started in Phase II

❖ INTRODUCTION

Although a lot of work has been done in Phase I, the needs assessment committee (NAC) has insufficient information for making needs-based recommendations to the organization. More is required, and most of the time it comes from surveys (Web-based or regular mail versions), further analysis of records, epidemiological studies, other related techniques, and qualitative sources (face-to-face or cyberspace focus group interviews, individual interviews, nominal group sessions, observations, and community group forums or meetings). How the NAC starts into Phase II is the central focus of this chapter. It begins by discussing

- the need for more information,
- what mechanisms would work best for attaining information,
- whether multiple methods should be used and how such would be pulled together,
- what funds are available for Phase II, and
- what activities might be most beneficial overall.

What additional information would be most useful for the needs assessment? The answer flows out of the initial explorations and data-gathering efforts from Phase II. The NAC already knows a good deal from records, prior evaluations and assessments, census data and related sources, discussions with key actors in the local context, reports from regional planning commissions, business groups, and state and national agencies. The results of Phase II are summarized in tables of what has been learned to date. Quickly review them—their content is a foundation for Phase II. They should be examined by the NAC before it arrives at a consensus that more is needed for making decisions about needs and future directions for the organization. Since Phase II entails additional costs (some possibly large), the group must be in agreement about seeking more data.

What does the review produce? The group may say it has looked at issues from different perspectives and there are no apparent problems that necessitate further attention. The organization does not have to do anything else, and the process has come to a logical end. Why should precious resources be allocated for a very limited payoff that would outweigh any potential benefits? The needs assessment may therefore be concluded.

A second option is that the NAC has a good grasp of the underlying problem(s), what is (are) causing it (them), and even what some possible solution strategies might be. Because of this, it can go directly to the end of Phase II or to Phase III and begin to think about solutions and plans for resolving problems.

Being able to move from Phase I directly to Phase III is desirable and will have a positive impact on the needs assessment budget. Many assessments can do this, thus bypassing the collection of additional data especially when not much value would be gained. This saves time and helps in maintaining the interest of the NAC. Long assessments can dissipate motivation.

The option of generating more information from new methods comes with costs from monetary, motivational, and schedule standpoints. But the NAC may be uneasy, and there may be political issues that require a stronger base of data for making recommendations. The committee wants a deeper sense of the need and what might be important to various constituent groups.

❖ MAKING THE PHASE II DECISION

Those summary tables (1.1 and 1.2) from Phase I illustrate the value of previous work and offer much guidance for Phase II deliberations. Ask probing questions such as the following:

- What are the understandings we have gained from Phase I?

- Are we at a point of agreement?

- Is the information sufficient for making Phase II or Phase III decisions or for terminating the needs assessment process, not going further?

- Do we feel comfortable with the decisions we are proposing?

- Are prior results in enough depth to justify our decision to the higher-level groups within the organization?

- Are the results corroborative, or are there any major sources of contradictory evidence?

- Are there missing areas that call for the collection of more data?

- Are these gaps of enough concern to propel us into Phase II?

- What methods might help fill in the gaps?

- Do we know enough about what different levels (service recipients, providers, the overall system) think about the needs areas?

- Are there other facts or opinions that should be gathered for enhancing the picture of needs?

- Would Phase II be worth it, especially if the costs are high?

- Which Phase II activities would yield the most useful information?

- What data sources would appeal (have credibility in a persuasive, positive way, not a political or negative one) to the organization?

- How could we pull together what might be a more complex, mixed methodology data set?

Table 1.1 One Useful Format for Displaying the Initial Work of the NAC

Area of Concern	What Should Be	What Is	Sources of Information	What We'd Like to Know	Sources of Information
Area 1 Subarea Subarea	Standards, expectations	Current status	Records, archives	More about status, perceptions of status, etc.	Other records, interviews, etc.
Area 2 Subarea Subarea Subarea					
Area n					

Source: From *The Needs Assessor's Handbook*, by J. W. Altschuld and D. D. Kumar, 2009, Thousand Oaks, CA: Sage. Used with permission.

Table 1.2   A Phase I Decision-Oriented Framework

Need Area and Subareas	Further Action Required	Reasons for Further Action	Preliminary Ideas About Causes and Solutions	
Area 1 Subarea 1 Subarea 2				
Area 2				
Area 3				
Area n				

Source: From *The Needs Assessor's Handbook,* by J. W. Altschuld and D. D. Kumar, 2009, Thousand Oaks, CA: Sage. Used with permission.

The list gives a feel for the discussion of the NAC. The facilitator does not discourage the NAC from implementing Phase II activities, but the group should not prematurely jump into them. Deliberation and exchange of ideas are warranted. If the choice is to go forward, what methods should be considered?

❖ WHAT METHODS WOULD WORK BEST FOR ATTAINING INFORMATION?

It depends on the ability of the facilitator, the resources available for the work and the use of external assistance if necessary, the skills of the group to conduct a variety of methods or to work with others conducting them, the time required for implementation and if it will allow for the collection of meaningful information, the face validity of the methods within the organization, and whether the data from additional methods would expand or support prior understandings. What frequently happens is that a quantitative technique (a survey or an epidemiological study) and a qualitative technique are seen as appropriate. There are logical reasons why these multiple methods are used in Phase II and work well together.

Some needs (social problems, educational deficits, health care issues, poverty, community relations, drug and substance abuse, violence, worker motivation) are complex and cannot be fully understood through the lens of a single method and its philosophical perspective. The idea is to look at the concern from multiple viewpoints and angles—to see the picture in different ways and to broaden the horizon from which it is contemplated.

Additionally, remember that Phase II of needs assessment relies heavily on records and archival data. Databases are examined, past studies perused, literature sources scanned for insights they might reveal about the problem and how it was studied (variables, methods), and reports from government and/or private agencies sought for guidance about the current need and features related to it. That is good, but Phase I strategies tend to lack direct input from service recipients and providers and the organization supporting them.

The last point has subtle dimensions—past versus more current data and collecting the views of the three levels of need. Most sources in Phase I are archival and may be dated. As an example, information regarding childhood diabetes (as derived from the National Health and Nutrition Examination Survey, NHANES) indicates that over a short period of time the number of children in the United States who are overweight or at risk for being overweight rose by 1%–2% (Chiasera, 2005). If the trend is based on a longer period, the rate of change might be lower because some of the older data might not be of as high a quality and would not fully reflect what is currently taking place. The childhood diabetes problem and its effects on health may be underestimated due to datedness.

On the other hand, the NHANES as a data system shows the strength of having information from multiple methods. While most of it is about physical measurements (height, weight, blood pressure and glucose levels, body-mass index, and other standard health measures) and demographics (gender, ethnicity), there are also data from intensive interviews about lifestyles, eating patterns, and so forth. The combination of methods is dramatic and, in totality, enhances understanding about the disease, not only what is going on now but what might happen in the future and what in the lifestyles of children might contribute to a possible epidemic at a later time. For diabetes, this is as meaningful as the "hard" health data in learning about the rapid rise of the disease. Each source does not give as complete a picture of need as when all of the sources are played off each other.

Why do quantitative (surveys or other techniques) and qualitative methods seem to be so prevalent and routine in the practice of needs assessment? Notice it isn't two quantitative or two qualitative methods but one of each. In years of teaching program evaluation, the author strongly advocated the mixing of methods in just this way from his evaluation and assessment experience. The wisdom goes something like this:

> Numbers are concrete and believable, and if not presented, some decision makers will discount what the assessment has produced. Basic descriptive statistics are common in the newspaper and popular literature. They are part of everyday life. In needs assessment, they show in a straightforward manner what the discrepancies are or what the situation is. Without them we cannot carry the day.
>
> But they are just numbers and by themselves cannot represent what it means to go through a health crisis or to experience the consequences of poverty. (What is it like to be poor or live in poverty? How does it affect one's self-image, especially if a person does not have a job or is homeless?)
>
> Numbers may not be able to adequately inform us of the recreational, arts, or social needs of a community. We need to go beyond them for meaning, and that is an aspect of the patchwork quilt of needs assessment that qualitative data can provide. When we combine the two types of methods, believability in the results may rise exponentially. In many cases, information from qualitative sources helps decision makers "feel" the data—which come alive from the qualitative results (e.g., from case studies) that demonstrate what the quantitative data represent in terms of human lives. Example 1.1 is a demonstration of this principle.

Example 1.1

Health Care From a Personal Perspective

The author's secretary was caring for her aged father with limited family or other assistance. As his need for care intensified and required major intervention, she sought help from Medicare, Eldercare, his pension system, her employment health care program, university-based specialized programs, social service agencies, and similar providers. Since her office was right outside the author's, unintentionally he heard snippets of the phone conversations about the situation. Just the amount of time to go through all of the different providers was staggering.

What emerged was that the health care system (or is that *systems?*) was a complicated jigsaw puzzle of many players with varied roles to play. There were all sorts of eligibility, income thresholds, and other criteria to be met. It was like trying to weave through a complex, difficult, and hard-to-navigate maze with many turns that were, in reality, dead ends. In a sympathetic vein, one could sense her exasperation and discouragement about the experience and the toll it exacted.

Quantitative procedures for understanding health care needs are certainly useful and required, yet they seem to be inadequate to describe what is taking place from a personal point of view. The effect not only on the person in need but on the caregiver and loved one is apparent. This problem occurs in other countries. Marklund (2004) wove it into the thread of a mystery. Perhaps that Swedish writer went through the same type of experience as the secretary did.

Observations, focus group and individual interviews with varying levels of staff and recipients of services, and other such studies are useful. They reveal the human side of the equation and the despair that might (*will* is a better word) never show up in the statistics. Statistical presentations cannot convey these dimensions. To her credit, the secretary was doggedly persistent and eventually successful in pursuing options. Others may have simply given up in disgust and quit. Both qualitative and quantitative data alter and shape understanding and interpretation and are necessary for needs assessment.

Before going further, there is a subtle problem with archived data that must be raised. Quantitatively oriented assessors are sometimes enamored with numbers and may be unquestioning about numeric information. There is no doubt that analysis and use of quantitative results are important for needs assessment, but stop and ask how the data were entered into the base in the first place. What was included or excluded? Is there anything in the entry process that might affect interpretation or that should be thought of when looking at analyses?

There are many illustrations of what could happen by just blindly accepting the validity of data. In education, for example, there may be legal issues associated with teachers putting into writing concerns about a student and their perceptions of behavior. Could they be sued? Are their complaints or issues open to inspection by administrators or others? Would they receive system support if sued?

Might such concerns constrain what they enter into formal, open-ended comments? They may not state what they are really thinking. Other similar illustrations can be found in health (up-coding of illness to receive higher remuneration from an insurer; one illness in medical records taking precedent over another, especially a physical one as compared to a mental one) or in economics (inaccuracies in the consumer price index) noted in prior needs assessment literature (Altschuld & Witkin, 2000).

Therefore the NAC and others working on the assessment are encouraged to probe into how a base was created and how data were entered into it. Do not take the meaning and quality of archived data on faith. This idea emphasizes the stress placed on having quantitative and qualitative data in needs assessment. More viewpoints on a problem are desirable.

❖ SHOULD MULTIPLE METHODS BE EMPLOYED?

The answer seems obvious, but the NAC and the facilitator of the assessment process must be clear as to what is involved in multiple methods. Do they possess the expertise to interpret and pull together all types of information into a meaningful, focused summary? What might happen if there is a *within-method* variation (explained later)?

The trade-off seems to be to use one method and get less information or to use more than one and increase the explanatory burden in a major way. The picture is more involved when information from different groups is obtained not from the same method but from different ones. Altschuld and Witkin (2000) described such an assessment in health and the difficulties observed in combining quality data coming from quite divergent sources. Arriving at a coherent synthesis was difficult in this instance. The decision about multiple methods should be entered into not lightly but only after careful planning for ways in which one source would complement or provide additional information not generated by another.

In terms of qualitative approaches, determining what the data tell you may be more difficult in practice than anticipated. Deriving themes across respondents, groups, and questions may be elusive. Sometimes there may be multiple groups (focus group interviews with parents, teachers, students, and administrators in education) with somewhat different questions for each constituency. Thus, within the qualitative technique, another layer of interpretation is being added to the data.

An extensive grounding in social science and/or the specific setting of the assessment is an important quality to have on the NAC for bringing out meaning. Beginners often have trouble in interpretation, especially when it comes to identifying explanatory themes that connect sections of the data. Experience plays a major role in extracting themes. Choose the members of the NAC wisely for what might be entailed in the analysis of qualitative results. What might their insights be able to contribute to the understanding of the data?

The same logic would apply to dealing with quantitative methods when the NAC has to draw statistically valid conclusions from them. Some results are straightforward such as percentages, frequency counts, descriptive tables, bar charts, pie diagrams, and the like. Similarly, inferential statistics (testing hypotheses, generalizing beyond the sample that provided the initial data) can range from the basic to the complicated.

This may occur when we move from statistics to the formulation of policy. Consider the case of raising the standards on school tests and what effects might be observed by doing so. What are the implications of changing pass and fail points for schools, teachers, students, and programs? What might be the political fallout from lowering the fail point on graduation rates and perceptions of the school system? Are there possibilities for "gaming" the system? What do statistics tell us about the status of health in the United States (or another country), and how do we translate them into utilitarian policies? Or, in the business and industry sector, what might happen if we dramatically raised the fleet mileage standard for automobile manufacturers?

Some of these questions take us into epidemiology (see Chapter 4) and futures forecasting. The methodology and the rationale underlying projecting become complicated, depending on how far into the future assumptions about it are made. We probably would be able to estimate the immediate, short-term impact on pass-fail rates, but a lot of other factors could affect the estimates.

What might be changing in school systems to alter test scores? What resources are needed to deliver new instructional strategies? How might the curriculum be restructured, or should it be restructured? How might teachers and administrators feel about pressures to improve scores? Would some subpopulations be affected adversely, more so than others? Could strategies be quickly shifted and resources directed to ensure that weaker students receive more attention, and what could be the impact of not emphasizing the higher levels of skills required for gifted students? (One argument against school standards and tests to assess them is the focus on individuals just missing the proficiency cut score and away from the gifted and/or very poor students.) Should we teach more to the test, and if so, might there be a negative impact on student creativity and interest?

Thinking in terms of Level 2 service providers (teachers), what are the potential positives and negatives for them of changing how things are done? Would they be accepting and motivated? It does not take much imagination to adapt this kind of thinking to business, government, higher education, and social services and programs.

For the futures forecasting, how good are the projections, and over what period of time would they hold? What assumptions have been made to determine the picture of some distant time ahead? What would other assumptions lead to, and what are the consequences of seriously flawed projections?

Sometimes, neither the NAC nor its facilitator has the skills and understanding to implement methods and analyze their results. The budget for the needs assessment should have a safety feature for hiring those who do as such skills and understanding might be needed for some quantitative and qualitative Phase II methods. Convincing administrators of this is necessary. To paraphrase the bard, "NAC and Facilitator, know thy skills well and be honest in your appraisal of them."

❖ HOW SHOULD THE RESULTS FROM MULTIPLE METHODS BE PUT TOGETHER?

The concern is that the data from Phase II activities and those brought forward from Phase I have to be integrated into a meaningful whole as much as possible. This is not easily accomplished. Analysis is taught but mostly from the isolated perspective of one type of method. Students learn how to treat data (qualitative and quantitative) but seldom how to think about evidence from multiple sets of data. In Book 4 of this KIT, details for how to pull data together are given. A few suggestions are offered in Table 1.3.

As you go down the steps, crossing out findings where there is agreement across sources, the lists per source rapidly shorten. The steps are a funnel with the number of items in Steps 8 and 9 being less, and better yet there may be none there. The steps do not deal with the quality of methods, but that is fairly easy to assess. If there was a well-conducted survey with high return and item completion rates and reliability, it would be stronger and take precedence over several quick individual interviews done for just some insight into the needs assessment situation.

Along such lines, if three to four focus groups were undertaken and revealed a consistent pattern of results, their quality would be apparent, and they would be of major importance for the assessment. The NAC should judge the adequacy of how each method was implemented and its validity (not in the strict measurement sense). This is a good idea, but don't devote a lot of time to making such assessments.

Table 1.3 General Steps for Handling Needs Assessment Data From Mixed Methods

Step	Description
1. Scan the results	Look at each method and its purpose in the needs assessment (ascertaining ratings of need areas, gaining perceptions of potential concerns held by groups, generating ideas, getting feedback about needs and the process, level of skills held by the group being studied, and desirable levels of same).
2. Observe main findings and patterns	Determine the main findings and discernable patterns from each method.
3. Array findings in order	Display the findings from the strongest (most supported by data) to the weakest per method. Strength comes from the numbers agreeing on an item, comments frequently stated in interviews, etc.
4. Observe agreement	Determine areas where the results across methods are in agreement.
5. Show agreement	On a separate sheet list findings that hold across *all* methods.
6. Indicate majority findings	Note findings where a majority of sources agree and there are no contradictory ones from any method. Contradictions do not often occur, but when they do, they are difficult to reconcile.
7. Think about partial agreement and some contradiction	Repeat Step 6 but for findings where there is agreement from some sources with contradictory and/or negative findings from others.
8. Consider single findings	Show findings that stand alone (come from individual sources) and for which there is no corroboration.
9. Consider findings in disagreement	List the findings in which the sources are in disagreement.
10. Decide if views of data collection should be included	Summarize perceptions regarding the data. (Clearly label this section as the opinions of a single individual or group, if from the NAC. The facilitator and the NAC are participants in the process, and their views may be valuable for decision making.)

If sources are corroborative, even if some were not well implemented, quality may be of diminished import as to its impact on the needs assessment results. With agreed-upon needs in front of the NAC, the dinner table has been set for use of the results (prioritization and action planning). Usually, there are too many needs for the organization to attack simultaneously. Some will have to be selected and their causes examined.

So far the chapter has been a start into Phase II with an emphasis on the need to obtain more information. Now a concrete choice must be made as to what specific methods should be described in subsequent chapters. This is not a simple decision since the number is relatively large. Moseley and Heaney (1994) cited numerous methods used across disciplines. In Book 1 of this KIT, Altschuld and Kumar (2009) presented an expansive list. Not all methods can be dealt with here, and therefore surveys, epidemiology, and a variety of interviewing procedures will be explained in detail. They are common in needs assessment with surveys being prominent, epidemiology being widely seen in some fields (health, insurance), and almost all assessments employing focus group or individual interviews.

❖ HOW MUCH BUDGET IS AVAILABLE FOR PHASE II?

Why wasn't this section placed earlier in the chapter? Doesn't the budget determine everything? Doesn't it put constraints on what we do or should consider doing? Why get worked up about things that cannot be done because the resources aren't there? If you begin with the budget, you put on blinders and may not be creative and entrepreneurial in thinking about methods. Why be hamstrung at the start before getting into Phase II? At times, there may be shortcuts that reduce costs and still yield a great deal of information. Think of ways to collect the data first, and if they are too costly, look for alternatives or other ways to implement them. Brainstorm and be open to all kinds of methods before getting into what the budget allows or doesn't. Ask what additional information you need and how you can get it. At the onset, there are no "right" or "wrong" ideas. All possibilities should be considered.

Budget might limit not what methods can be used but just their scope. While three or four focus group interviews (FGIs) might be desirable, only two might be done to save dollars. The same type of thinking applies to individual interviews or the number of observations to be made. Or some FGIs may be done in cyberspace (see Chapter 5) with less demand on scarce resources. There are many ways to implement or improvise with methods striking a balance between budget and practicality.

The sample for surveys could be limited in size or the number of groups involved. You could decide to do a survey on the Web rather than through regular mail. It would speed up return but would incur some costs (mostly minor) and would come with a few disadvantages (limitations of format, and only those with Web access or computer savvy could participate). Could the needs assessment accommodate the loss of some sample from not getting to certain segments of the population? Might some bias in the data occur with the use of technology? What are the trade-offs that the NAC is willing to make? Can some additional error be tolerated by not having ideal conditions? It may be that the confidence interval is larger, but what's the big deal? And with multiple methods, error might be compensated for if methods are corroborative.

Also remember that the NAC is a working committee with good leadership and carefully selected members. It could be a valuable resource for data collection. Could members of the committee be trained to conduct individual interviews? Could this be done in teams to cut down on bias that might creep into the procedure? Could the same be done for FGIs, and/or could surveys be distributed to intact groups? The only caveat is that it might be easier for the NAC to implement some techniques (e.g., interviews) than to analyze and interpret results.

Experienced facilitators teach the NAC to ensure quality, consistency, and objectivity in its work. Bias (probably not intentional) or some subconscious cueing could drift into an FGI or an interview without some standardization. NAC members have invested psychologically in the needs assessment, have given their precious time to it, and now have a sense of ownership. They may be stakeholders who have perceptions and expectations or have developed almost hidden anticipations for what might result. Without realizing it, they could in subtle ways influence interviewees as to "correct" answers even though interviewers should not promote such responses.

With brief training on how to interview, they would conduct the method in a more appropriate manner. This would tend to ameliorate problems. If small teams of NAC members were used, they could debrief after implementation and ask each other openly and frankly about such concerns, and this self-checking mechanism is encouraged.

Utilizing the NAC costs less than seeking outside assistance. There are limits to this, especially when NAC members have other demanding duties. When they can be used, a lot can be accomplished. When they cannot and funds are available, use outside investigators who have no attachment to the needs assessment and its area of focus. In this case it is beneficial to have the NAC participate with them to get a feel for the methods and what the data might mean. And even with experts you could still go with smaller samples or adapt methods to cut expenditures.

As an illustration, instead of doing a standard Delphi survey (a technique with many iterative surveys with each survey being adapted from the prior one), why not distribute surveys on the Internet (Hung, Altschuld, & Lee, 2008) or use the Group Delphi form of the procedure? The latter utilizes an intact group with the surveys being completed in one 3- to 5-hour session. It is not the same as the standard procedure, but it might work as well.

For such options, make sure that the strengths and weaknesses (what information you might not get or might get less of) are clear in terms of the possible effect on needs assessment outcomes. Most facilitators are trained in methodology and know the pros and cons of methods. Many times the loss from alternatives is tolerable and acceptable. If the facilitator is unfamiliar with a method, a quick search of the literature or the Internet and/or outside advice will often provide the necessary insights.

❖ OTHER RESOURCE OPTIONS

If the organization has an office of institutional research (evaluation or planning), see if you can use it. Nearby universities might be contacted about graduate or advanced undergraduate student help in exchange for the experience of working on a needs assessment. Many students in relevant fields seldom have been involved in such an assessment, so if they are, make sure their contribution to the organization and the effort is noted in final reports. Other agencies (state offices, regional planning groups, etc.) might also be willing to assist.

The final thought is one that will give headaches to administrators when they review needs assessment budgets. Due to many unexpected twists and turns, it is almost impossible to specify what will happen in Phase II when the assessment starts in Phase I. There is no template that permits accurate prediction of what will occur. The need for specialized methods and skills may arise. The prudent facilitator anticipates this and includes some funds for it in the initial budget (or finds qualified individuals and solicits free assistance). Such funds should be noted in the first budget, and administrators should be alerted that more funds might be needed at certain times. The facilitator's knowledge of the organization (see the Cultural Audit in Book 2 of the KIT) is helpful in approaching this delicately and tactfully. Maintaining channels of communication and using them is of major importance.

❖ TOWARD A PHASE II SET OF ACTIVITIES

Deliberations are now completed, and they point to the need for more information by qualitative and quantitative means. The committee pursues this course with full knowledge of what might be involved. Everyone is primed for the arduous yet potentially very rewarding work of Phase II. In the next few chapters, more details are given about methods with examples of what others have done. Advice will be offered as to how procedures might be adapted to other contexts as well as pitfalls to avoid. Needs assessment is characterized by subtle judgments made within the parameters of unique situations shaped by political circumstances, groups, and individuals. As necessary, make adaptations.

Highlights of the Chapter

1. Don't rush into Phase II activities; they require new data collection methods, analysis, and interpretation strategies.

2. Careful deliberations are required by the NAC given the additional costs and whether the new information will be of value for needs assessment decisions. Build from Phase I as much as possible; the needs assessment process overlaps and is interrelated.

3. Generally, the idea of using multiple methods with one being qualitative and the other quantitative is stressed for understanding complex needs. This is a good feature of needs assessment and is strongly encouraged.

4. Multiple sources of data and information from Phases I and II have to be amalgamated into a meaningful whole for decision making. This is not easy. An ordered strategy for doing this was offered.

5. Costs for Phase II could be the "Catch-22" or the fly in the ointment. In anticipation of how funding may impact plans, ideas were given for getting around the potential roadblocks due to insufficient resources.

6. Lastly, communication is an essential ingredient in a successful needs assessment. Never lose sight of that fact and its relevance to the process.

2

Building From Phase I and the Literature

❖ INTRODUCTION

Phase I influences what takes place in Phase II. Tables 1.1 and 1.2 in Chapter 1 are two of the many summaries available to the needs assessment committee (NAC). There are more critical pieces to add to the mix of resources. In this chapter the emphasis is on them and using the literature to guide what is done in Phase II.

❖ USING PRIOR WORK

Witkin and Altschuld (1995) described a procedure that an NAC might complete early in Phase I (Book 1 of the KIT). The intent is to get the group thinking about questions and concerns that the organization might have about a goal or an area of interest. The facilitator kicks off a brainstorming session by stressing a probing, issue-oriented exploration of an area or a topic. Usually some brief and thought-provoking (to prime the pump) background is given. Attention-getting facts (on a page or two in bullet fashion) about the current state of affairs are a good procedure to jump-start things.

The group first looks at the goal or area and raises issues that would be useful to understand from a needs assessment perspective. From there the process moves into ideas for obtaining data about the issues including sophisticated and quick and dirty methods. The table that is eventually produced is utilitarian and may often be referred to in the work of Phase II. Because it is brainstorming, the more ideas generated, the better. The NAC should be challenged to be expansive. The activity doesn't take much time and is rather fun. Questions that could be used are:

- What do we know about the recipients of services?

- Where are they currently in terms of awareness?

- What opinions do they hold?

- What are their attitudes?

- What are their behaviors in terms of skills and actions?

- What is the desirable state (that which ought to be) for some of the above questions?

- What do we currently understand about all of this?

- What answers would we find of value for needs assessment deliberations?

- What sources of information exist in these regards?

- Where are they located, and how difficult or easy is it to access them?

- Do the sources contain in-depth information about the questions?

- Are there places where we will have to create the data for answers?

- Are there quick and simple sources that are inexpensive to get information from, before others are considered? (Caution is in order since quick and simple is good, but such criteria could curtail brainstorming.)

It is helpful to provide a sample of a completed table for perusal and commentary. (Table 2.1 is one such example, and Table 2.2 is a blank form.) The facilitator emphasizes that once finished a table is a general map for what might be done in an assessment, not that everything in it would be religiously followed. Table 2.1 shows what could result from a brainstorming activity. It should be noted that the data collection strategies are an interesting mixture of quantitative and qualitative approaches and might not require much in the way of effort

Table 2.1 Data Resources List Format for Preassessment

Goal: To revise our curriculum in educational research, evaluation, and measurement			
Concern: What do we know about our students and why they come to our program? How does our curriculum match up with that of other institutions? What skills and knowledge are our students using in their work? What skills will be needed in the future?			
What Is Known		*Data to Gather*	
Facts	*Sources*	*Facts*	*Sources*

Past students
  Degree levels
  Gender
  Countries
  Current jobs
Courses
  What we teach
  How concepts relate
	Records
	Faculty notes
	Syllabi
	Syllabi review
	Group discussion
	Job opportunities
	Requests for services
		Complete listing of jobs held
		How training relates to current work
		Publications
		What do other curricula and courses look like?
		What do our competitors do better than we do?
		
		Opinions
		What do current students think of courses?
		What do past students perceive as important and/or missing?
		Why did they choose us?
		What are their expectations?
		What do other consumers (other faculty members) think of us?
			E-mail survey
			Collect current résumés
			Literature review
			Phone interviews of other universities
			Collect other syllabi and benchmarks
			
			Sources
			Focus group interviews
			Surveys
			Phone interviews

Source: From a presentation by J. W. Altschuld, 2003, Summer, Workshop for Korean Educators, School of Educational Policy and Leadership, The Ohio State University–Columbus.

Note: More columns may be added to indicate who will be responsible for gathering data and target dates.

Table 2.2   Data Resources List Format for Preassessment

Goal:			
Concern:			
What Is Known		*Data to Gather*	
Facts	*Sources*	*Facts*	*Sources*
		Opinions	**Sources**

Source: From *Planning and Conducting Needs Assessments: A Practical Guide,* by B. R. Witkin and J. W. Altschuld, 1995, Thousand Oaks, CA: Sage Publications. Used with permission.

Note: Additional columns may be added to indicate who will be responsible for gathering the data and target dates.

to implement. There is strength in combining methods, and some combinations might be readily apparent when the table is reviewed.

Returning to the process, based on the introduction, the group proceeds to identify a goal or topic area (the facilitator could do this in advance, if appropriate). Each person *individually* develops concerns that he or she has about the goal. The concept is not "Here are some data that we can access in a rapid manner." Instead it is "What is troubling us about this goal or area? What are some of those deeper underlying things we need to probe into for understanding of the concern?"

Continuing as individuals, NAC members fill out the form with their rough thoughts. Allot 15–30 min for the individual work depending on the topic and the skills of the members. The group reconvenes for an analysis and synthesis session. Concerns are reviewed and collated into a summarized list. There may be some differences, but there will be enough overlap to collapse most of them together. Lastly, the group completes a final table consisting of consolidated entries. Even if time is limited, much value can be accomplished.

If this form was used in Phase I, just return to it. It is a terrific roadmap, an outline of a plan that can be referred to periodically, and is a quick way to focus on what might go on in Phase II. Since the brainpower of the NAC was behind the finished table, ownership and buy-in should occur. By the same token, there will be some need to focus on some data collection methods over others due to resources, organizational timelines, cost, and the particular skills and talents of the NAC. Creating the table is a change of pace for the group.

❖   GUIDANCE FROM THE LITERATURE:
     SUBSTANTIVE MATTERS

The literature is critical for needs assessment. Reviewing what can be found prior to conducting the assessment study cannot be stressed enough. How others have conceptualized and conducted needs assessments for this or a related topic is instructional for the NAC and its facilitator. What methods did they use, what seemed to work well, what did not, what constituencies were involved, how did they analyze results, how did they reconcile differences that may have occurred across constituencies, how were the data portrayed, and what advice did they offer other needs assessments?

These are important topics mostly of a methodological nature, and while of value, they may be somewhat less so than the substantive focus of the assessment. It is fundamental that needs assessors go beyond techniques to find out about the major concern of the assessment. Methods are fine, but they do not substitute for a deep understanding of constructs and theories from previous studies.

In the entries in Table 2.3, the literature was used in this manner before starting the assessment process. This led to some clever features being imbedded in needs assessment designs and, in several cases, to the incorporation of theory-based items in instruments. As a result, the assessments were taken to a higher level of quality and produced more

Table 2.3 Overview of Multiple Methods Needs Assessment Studies

Author(s)/ Years	Focus	Variables	Populations (Groups)	Methods	Comments
Altschuld et al. (1997, Part 1) Cullen et al. (1997, Part 2)	A *retrospective* needs assessment imbedded in an evaluation of a national service in education An in-depth case study of selected sites related to the evaluation Corroborative evidence from multiple methods	Importance of key resources for improvement Availability of those resources Discrepancy scores obtained from importance and availability scores Aspects of how sites went about improving and/or implementing science education programs	National samples of teachers and administrators in U.S. schools (surveys) Administrators and teachers in a small sample of selected schools (case studies) National samples of service users	Surveys (Part 1) Case study (Part 2) Case study interviews in selected schools Comparison of results across groups and methods	Needs assessment double scaling used Complicated within-methods variation imbedded in surveys for teachers and administrators Not primarily a needs assessment but elements of it were included, especially in surveys Corroboration across methods led to important insights for needs Problems of vernacular noted in case study interviews leading to a subtle form of a within-method variation

Author(s)/ Years	Focus	Variables	Populations (Groups)	Methods	Comments
Holton, Bates, & Naquin (2000)	Training needs assessment Conducted within a large government agency Performance of the organization was emphasized	Definition of training needs by high organizational level particularly as related to performance Training and nontraining solutions from staff level Separation of wants from needs Especially heavy leadership weighting given for what constitutes performance for an organization	Multiple levels within the administration of the agency (really two levels within administration) Online staff but content different from that of the administration level Subject matter experts	Detailed, probing interviews Input from subject matter experts Surveys Synthesis of multiple sources of information Focus group interviews	Two different approaches were taken depending on the level (administrators and staff) in the organization Part of the strategy was predicated on reducing the emergence of "wants" in the assessment Use of multiple methods was apparent, especially interviews and an interesting variation of a presurvey for focus group participants Noticeable difficulty in defining major discrepancies on the part of those interviewed Organization was very large

(Continued)

Table 2.3 (Continued)

Author(s)/ Years	Focus	Variables	Populations (Groups)	Methods	Comments
Hunt et al. (2001)	School dropout and violence and prevention of same	Perceptions of factors related to dropping out and violence Interventions pertinent to prevention of the two problems	Students in the school system Teachers and administrators Parents Multiple school levels Others	Surveys Individual interviews Focus group interviews Interactions and discussions	Multiple methods are prominent Major involvement of different Levels 1, 2, and 3 in the process Mix of needs and solutions in the assessment to enhance ownership Prolonged engagement of needs assessors onsite Some within-method variation used
Chauvin, Anderson, & Bowdish (2001) Chauvin & Anderson (2003)	Training needs for public health preparedness Issues relative to the training identified Problems in public health delivery for disasters and other calamities	Areas of training needs clustered by categories in accord with national specifications Motivation to enhance levels of skills Preferred or desired training modes Barriers to training	A variety of levels within the public health system - high-level administrators - mid-level administrators - line/direct delivery personnel	Surveys Survey items based on available literature Much detailed work in survey including the involvement of experts in initial review of items	Triple scales used Interesting approach to analysis Unique way of displaying the results and that in turn may affect their interpretations Good blend of scaled and open-ended questions on surveys

Author(s)/ Years	Focus	Variables	Populations (Groups)	Methods	Comments
Altschuld, White, & Lee (2006) Lee, Altschuld, & White (2007a, 2007b)	Minority student retention in a statewide alliance of universities Part of a nationwide set of alliances Needs assessment inside of an evaluation	Importance and satisfaction of services for students Frequency of their use Discrepancies in skills being learned and satisfaction with same Open-ended reasons for different responses to questions from groups in the study	Students in universities in the alliance First-year students to seniors Faculty in the universities Administrators/ others who implement retention services (incorporated in the above group on surveys)	Student surveys (scaled) Faculty/ administrator surveys (scaled) Open-ended follow-up survey sent to both groups Web-based surveys	Triple-scaled surveys Within-method variation used Discrepancy analysis for importance of and satisfaction with services Frequency-of-use scores were relayed to decision makers but not included in discrepancies Follow-up looked at reasons behind individual group responses and why they differed between groups Emphasis on some of the unique data problems encountered with triple-scaled surveys

meaningful information. The time to locate and read sources depends on budget and staff resources. When such are not plentiful, dividends will be realized even from a quick search and a skimming on the surface of journal articles and reports found online.

❖ GUIDANCE FROM THE LITERATURE: METHODOLOGY

A number of excellent multiple methods assessments are described in Table 2.3. They provide direction for conducting Phase II, from conceptualization to implementation, and have many of the basic characteristics of needs assessment being promoted throughout the books in the KIT. They employ qualitative and quantitative methods and exhibit the deliberations that make for successful yet somewhat more complex assessments. The price of complexity is offset by what these assessments accomplished. Some of them have what is called a "within-method variation," a variation of a method to fit different groups in the needs assessment environment. The studies afford a set of guidelines, empirically based, for doing a good job of gathering assessment data. Many of the activities will fit other settings.

Each row entry (for some, several studies were carried out) is summarized in regard to focus of the assessment, types of variables assessed, groups sampled, aspects of the methodology, and other comments. From there it is a simple matter to extrapolate principles for planning assessments and to use the later parts in the book.

Before proceeding to design principles, only one of the studies from the literature included a description of a formal NAC. That doesn't mean that it is not necessary and that the rationale given for it so strongly in Book 1 should be disregarded. The one study (both an assessment and an evaluation) utilized an external board that participated in project work and established policies and parameters for what was done. The final report complete with recommendations was done by the board, which for all purposes was a formally constituted NAC. Five of its six members were from states other than the one in which the organization was located. The local member served as the facilitator on behalf of the other five. The facilitator drafted general guidelines for the work, which were then intensively reviewed by the board before anything was undertaken. Background information à la Phase I was provided to the board. This was fairly easy to do because the organization was only 2 years old and extensive data were not available. Lastly, the organization provided staff and resources for the project, but the effort was under the auspice of the board.

For the other studies, guidance for the needs assessment is there but in an understated way rather than having an NAC in place. There were many steps in the assessment process where outside individuals were consulted or where project personnel were in frequent and close communication with internal staff and key stakeholders. A quiet, NAC type of steady hand was ubiquitous, but a formal group was not.

❖   PRINCIPLE 1: CAREFULLY SELECT
     CONTENT FOR THE NEEDS ASSESSMENT

In all of the studies, much time was spent on the content of surveys, interviews, and so forth. Ideas—understandings of what was desired in the local situation and what it would tolerate—came from the literature, and in several cases, theory played a major role in instrument development. In one instance, the authors postulated that the literature enhanced the quality of the needs assessment and the interactions with Levels 1 and 2 in the context.

❖   PRINCIPLE 2: INCLUDE LEVELS 1, 2, AND 3,
     IF POSSIBLE (ALSO SEE PRINCIPLE 4)

The studies mostly focused on at least two of the levels. Service recipients and providers were part of the process. Sometimes the service providers (Level 2) were subdivided into those who were directly involved with Level 1 and administrators (Level 3) who were close to the situation. In one instance, two levels were involved but with a pronounced weighting dimension; responses from Level 3 were given greater credence than those from Level 2.

Across most of the sources, Level 3 or the system did not appear to be prominent in Phase II. It may be that it is more pertinent to making decisions about needs and planning and implementing solution strategies than to the collection of some of the basic data. Hence Level 3 would not be seen as much in Phase II. Another possibility is that concerns relative to supporting the provision of services for Level 1 (service recipients) automatically arise when information is sought from Level 2. What Level 3 might have to do is there by a process of osmosis rather than by direct involvement.

Sampling was carefully carried out for the groups and individuals included in Phase II such as how to (a) reach different groups, collect their data, and deal with results if they are in conflict; (b) use intact groups as staff in a business or students in classes; (c) work with groups that are geographically dispersed by collecting responses via electronic means; and (d) think about situations in which some groups might be more likely to respond by e-mail whereas others might like the tactile feel of a survey. The needs assessment picture becomes more complex if different methods are employed for varied groups. Small focus group interviews may be better for young students to find out about needs whereas surveys may be sent to parents and individual interviews may be conducted with teachers. Trying to construct meaning across methods can be difficult.

❖ PRINCIPLE 3: EMPLOY MULTIPLE METHODS

A notable feature of all of the studies was the use of a quantitative method (mainly the survey) with a qualitative one (interviews, focus group interviews). This is more expensive and complicated since integrating and interpreting data from two or more sources is time-intensive and costly. For the needs assessors whose work is cited, the additional resources were outweighed by the payoff in understanding.

(This is in contrast to Witkin's 1994 research and informal investigations in the author's classes over the years in which single-method needs assessments constituted the vast majority of what was done. In them, determining discrepancies or even a sense of discrepancy was generally absent, leading to the observation that many activities labeled as needs assessment might not be such. Hopefully a subtle shift in the state of the art is occurring.)

❖ PRINCIPLE 4: VARY INSTRUMENTATION
TO FIT THE SUBTLE DIMENSIONS OF MULTIPLE
GROUPS (WITHIN-METHOD VARIATION)

One study had a planned *within-method variation*. The same or similar information was wanted from two groups, but there were disparities in the way in which they would view the problem area, so the same instrument would simply not work. The wording and order of questions in surveys had to be tailored to such distinctive qualities.

In yet another study, the ages of respondents necessitated alternative wording of questions. Adult queries had to be placed into the vernacular of children. The same kind of thing happened in the study guided by the external board where teachers were not familiar with many systemwide initiatives, and the initial framing of interviews had to be changed accordingly. Questions for administrators simply did not resonate with the classroom direct-service-delivery level.

An interesting way to think about a within-method variation was presented by one needs assessment. Because unique purposes were assigned to the inclusion of Levels 3 and 2 in it, respondents received different questions. It must be noted that when this is done, the interpretation of results becomes more difficult. If, for example, three groups in an assessment received similar yet somewhat altered questions in varying orders, could the same meanings be ascribed to the results? Did the respondents see the questions in a way that would allow us to compare the data? Could order, wording, and/or combinations thereof affect outcomes? Within-method variations are necessary, but there are no easy solutions to the subtle systematic bias or slant that they may interject into the data.

❖ PRINCIPLE 5: CONSIDER TWO OR MORE SCALES FOR THE NEEDS ASSESSMENT SURVEY

When a discrepancy definition is stressed, conducting "what is" and "what should be" comparisons is a reasonable way to proceed. (However, one group of authors questioned whether the idea of formal discrepancy works well in very large systems.) The two conditions or at least a sense of them is necessary for a need.

But by placing two or more scales on a survey, it may be harder for respondents to complete the instrument, particularly if they have limited familiarity with double (or more) scaled formats. Hamann (1997) found that when the number of scales was increased from one to two, the return rate was negatively impacted, and the finding equally applied to those who were more highly educated as well as those with less education, experience, and sophistication. As more scales are added, return and/or item completion rates may go down and affect results.

The problem is compounded considerably by the fact that item completion rates are usually higher for what should be (everyone has feelings, opinions, or perceptions) but generally lower (and even much lower) for the "what is" side of the equation. There could be different numbers of respondents for the two scales, and calculating the discrepancy score between them presents real concerns. Think what might happen if three or more scales were employed (Chapter 3) and how the problem would be compounded. Not all of these issues can be resolved. One partial solution would be to have three or more scales, but groups of respondents would answer surveys with only two of them. The different surveys would randomly be sent to respondents, and from their responses the results of the entire group on the three scales could be estimated. It's more complex statistically but worth a mention.

Another possible solution is to have response options such as DK (don't know) or NA (not applicable). The data have meaning and should be treated not as missing but as representing respondents' understanding of the issue. NA might denote that they haven't participated in an activity, and DK indicates that they really have no knowledge upon which to base their response. The latter often happens with questions about current status. In this regard, you could imbed short descriptions of an area prior to questions about it so all respondents would have some facts at hand and a common frame from which to answer. The survey would become longer (perhaps with less return), but conversely, respondents would be answering from the same base of knowledge.

The bottom line is what is the best way to structure the instrument? How many scales should be placed on it if there are so many problems as explained above? It depends on the emphasis and nature of the needs assessment being conducted. If two scales (or possibly more) would be good for understanding needs, use them.

In Table 2.3, one of the entries was especially notable in its use of a single survey with three scales for items clustered in sections. The sample consisted of service providers (Level 2) and administrators (Level 3). No within-method variation was necessary. Extensive demographic data were collected, enabling the breakdown of results by and facilitating comparisons across levels. Probing, open-ended questions were also included in the survey.

Because of the problems arising from multiple scales and return/ completion rates, encourage respondents to answer all questions due to the importance of the information for identifying needs and, in turn, improving situations. Stress the value of responses for guiding decisions. By attending to such details, the survey process works better, and fuller information is obtained. The author and his colleagues have done some research regarding survey use and analysis in needs assessment. Even with the challenges given above, they strongly recommend that multiple scales be part of the instrument because of the information yield. It expands thinking about needs and positively influences discussion by decision makers. The weakness in data due to multiple scales is balanced by the depth of thought engendered in deliberations.

In that vein, consider the kind of data generated from three scales and the results as to how they might affect how committees view needs. Suppose the survey contained scales about the relevance or importance of an item, the level of attainment (achievement or satisfaction with it), and the motivation to work on the problem area. Two basic outcomes that could occur are as follows:

- Importance is rated high, attainment low, and motivation high.

- Importance is rated high, attainment low, and motivation low.

The first is the result we would like to see. Importance and attainment are showing a discrepancy that requires attention. The indicators are pointing in the right direction. With motivation to resolve the gap being there, everything is aligned. The gap is clear, and there will be strong support to reduce it. From a needs assessment standpoint, this is good with direct implications for the organization.

The second outcome is more troubling. The gap is the same, but the motivation to resolve it is not. If the topic of concern is very important, how is a committee to work with the situation? What course of action would be best to pursue? Is more investigation necessary to determine why motivation is low? What is happening or has happened in the organization that led to this pattern? What incentives might be necessary to implement changes? This result forces the NAC to carefully consider options and may foster better policymaking.

Now, let's add a different twist to the three scales. An NAC decided to use three scales, importance and satisfaction scales coupled with frequency of use for services provided to individuals. Two outcomes that might be seen are:

- high importance, low satisfaction, and low frequency of use and

- high importance, low satisfaction, and high frequency of use.

These results underscore two unique paths for the organization. The first set relates to the quality of the service (as in the gap) and questions about its promotion. The second (the gap with high frequency of use) is puzzling and might require a qualitative exploration with perhaps interviews and observations. (See Chapter 3 for a closer look at this type of data.) Individual or focus group interviews with users and unobtrusive observations could be undertaken. Why are they using a service that has such a big discrepancy?

The examples depict the value of data from more than two scales, but decision making becomes more intricate. Needs may be complicated, and the choices relative to them can be quite demanding. The data might take an organization to major changes in normal ways of doing things or into uncharted territory. New resources might have to be obtained or internal ones shifted from one place to another in the organization, leading to negativity caused by disruptions, loss of support, and/or loss of control of one's turf. Positions of power can be dramatically altered. Tensions can rise, and there can be serious consequences for how an organization handles its needs. The discomfort from having to deal with additional data when choosing a course of action is a price to be paid for the commensurate gain in reasoning behind a decision that is hopefully a better one. Data that are more revealing about the nature of the problem are warranted.

❖ PRINCIPLE 6: CONSIDER PROCEDURES TO COLLECT INFORMATION ABOUT BARRIERS TO SOLUTIONS AND PREFERRED WAYS TO IMPROVE

On occasion, include questions on surveys or in other procedures that do not deal with the identification of discrepancies. For training needs, surveys often ask about preferred modes for the delivery of training, the costs associated with same, when and under what situations it would be best to receive training, and so forth (Chauvin, Anderson, & Bowdish, 2001). Or surveys might be used to learn about barriers that currently exist or might exist to implementing change. This seems to violate the idea that needs are the main focus, with solutions coming in only after needs have been identified, explored in terms of causes, and prioritized. But in needs assessment you may have only one chance to get information from a sample, and you have to maximize what can be gained from a single opportunity to work with a group.

The NAC may decide that is the case, and at times, it is difficult to separate thinking about needs from thinking about solutions. It begins early in the needs assessment process. Survey respondents and participants in qualitative procedures may already be operating in this manner and have solutions invading their mental processing. In surveys or interviews, there can be questions about solutions put near the end of the exercise so as not to confuse things. Start with a clear emphasis on needs and move to these other considerations toward the end of data collection. Otherwise, it might be that too little attention is given to needs for the sake of solutions.

Highlights of the Chapter

1. Emphasis was placed on using what was done before or could easily be done at the start of Phase I to select the content of and the methods that might be used in the assessment. From inception to conclusion, needs assessment is an intertwined, intermeshed type of undertaking.

2. Guidance comes from the literature for the substance of the assessment, key variables to study, and methods to use. Even quick reviews of what is available are helpful for local contexts.

3. A number of multiple methods needs assessments were summarized, and design principles were extracted. They provide excellent input for other assessments. Adopt/adapt what others have done.

4. There are hidden subtle aspects to using the principles. Making them fit with the specific context and milieu of the needs assessment will be necessary.

❖ TOWARD THE NEXT CHAPTERS

The design and implementation of a selected set of techniques for Phase II will be described. The principles will be incorporated into the procedures, but they are not rules; nor are they intended to be. That point is important. Every needs assessment has unique dimensions depending on its importance to the organization, interpersonal dimensions, the likelihood of change that might result from findings, the decisions that are made, the skills of the facilitator and members of the NAC, the demands of methodology, the available budget, and a host of other factors. Rules are not appropriate—they can become too limiting and hinder the assessment endeavor. Principles imply modification, molding properties to fit, or general guidelines within which to operate. Think about needs assessment in this way. Be open to new and thoughtful ways to use that lead to a successful assessment in your situation.

3

That Pesky Needs Assessment Survey

❖ ❖ ❖

❖ INTRODUCTION

The needs assessment committee (NAC) decides a survey is necessary for Phase II. Surveys are prevalent in needs assessment and fit most local contexts and requirements. This chapter is not a substitute for the numerous survey texts dealing with virtually all aspects of design and implementation. The focus here is on what the NAC should understand about survey development and use in assessing needs. The NAC can accomplish much by thinking about suggestions in this chapter and using good common sense. In addition, most facilitators are familiar with the technique and its applications.

What are the key parameters for structuring and implementing successful needs assessment surveys? The principles from literature in the previous chapter provide essential guidance for the survey (Table 3.1).

Table 3.1   Overview of Principles for Needs Assessment Survey Design

Principle	Comments
1. Select content for the survey	It goes without saying that this is the most important decision in regard to surveys. Make sure that you are clear from your Phase I work and what has been learned from the literature when deciding upon content.
2. Include Levels 1, 2, and possibly 3	Nearly every needs assessment will routinely survey Levels 1 and 2. For reasons explained in the text, Level 3 is less often included.
3. Use other methods with the survey	When designing and implementing the survey, it is usually wise to think about employing other methods (qualitative ones) to enhance understanding of the area of needs.
4. Generate, if needed, within-method variations of the survey	Frequently missed in needs assessment is that surveys for different levels may have to be tailored to each level. This entails looking closely at the wording and order of questions.
5. Use at least two scales for survey questions	Maximize information yield by having two or more scales per item and carefully consider the sophistication of the audience when choosing the format of the scale.
6. Perhaps include statements about impediments and solutions in the survey	You may have only one opportunity to conduct the survey, so it may be desirable to have questions about barriers and solutions on it. Be cautious in doing so as the main purpose is to look at and explore needs.
7. Complete the survey with a few open-ended questions	Well-worded, prompted, open-ended questions can be very useful for needs assessment but may require extensive time for analysis. Use them judiciously.

❖   PRINCIPLE 1: CAREFULLY SELECT
    CONTENT FOR THE NEEDS ASSESSMENT

What is the intent of the survey, and what content should be there to reflect it? What the need is about is where the process begins. What specific ideas and concepts would benefit the organization? What information would move the organization forward and fit its demand for data about important concerns? Here are some examples of what might be included:

- respondents' behaviors, what are they doing, and what they should be doing in terms of skills, safety behaviors, activities of importance, and so on;
- attitudes toward some kind of situation;
- perspectives about that situation;
- perceptions of what might be needed to rectify a problem;
- perspectives of what others might be doing or need to do;
- shorter, more current, and longer-term, future-oriented needs;
- satisfaction with what is currently taking place in the organization;
- importance and value of services being provided;
- degree of achievement with respect to skills;
- the motivation of individuals to resolve problems or to take action;
- perceptions of the willingness of the organization to change;
- difficulties or issues the organization is facing;
- frequency with which services are used;
- how often some activities are performed;
- feasibility of resolving problems;
- barriers or impediments to offering/using services;
- ideas about solution strategies;
- theoretical issues as observed in the literature; and
- what aspects of the situation people are interested in and willing to describe.

Some areas of content may not be about discrepancies or gaps but deal with the environment and problems in it. Many of the above points have implications for the format and structure of needs assessment survey items. Additionally, while most items will be scaled, usually a small number of open-ended ones will be used. Example 3.1 is a description of one study that used open-ended and scaled surveys for current and future needs.

Example 3.1

What Can Be Done With Surveys?

Looking at shorter- and longer-term needs can be more involved than it appears at first glance. A scaled survey for an engineering field had two parts—short-term technical training needs that were currently seen and technical training needs that might have occurred 3 years or more into the future with the latter being more speculative. The items came from an analysis of open-ended responses provided by a broad spectrum of engineers in companies in the United States. They were asked to identify immediate training problems and ones that might appear or be increasing with the passage of time. (Using an open-ended survey to guide the development of scaled ones as in this case is a form of the Delphi technique.)

A small group of engineers housed in a national center helped to analyze and categorize open-ended results. This was good since the group began to think more deeply about needs, how the field was changing, and issues in delivering "cutting-edge" training. Consider having technical (subject matter) experts assist in the analysis. It leads to a healthy and lively exchange and may produce better syntheses of data.

The statements derived from the open-ended responses became the basis of the two scaled parts of the survey. The current- and future-needs focus made the needs assessment more complicated and increased the work, but the information had greater potential for an impact on organizational thinking, discourse, and decision making.

The scaled instrument was sent to engineers in middle management from a sample of mainly large companies. They examined the items, selected their top seven immediate and future needs, and then rank-ordered them. The highest-ranking items from each list were noted as well as how frequently they were chosen for ranking. The ranks and frequencies of choice are valuable pieces of information. The instrument demanded more thought by respondents, but because the goal was to identify the most likely training needs for a very important national industry, the strategy seemed reasonable. As the process unfolded, the needs assessors found that they had to go to back to the literature more than anticipated for help in interpretation of results.

The national organization that commissioned the needs assessment wanted to develop training packages for the short term while considering what it might do in the longer haul. For the latter it could seize the initiative for future training, improve its leadership position in the country, and generate new business. The content of this assessment was obviously driven by organizational concerns. This was a needs-sensing activity, not needs assessment, since discrepancies were not actually obtained. On the other hand, what was learned in the assessment was benchmarked against trends in the field looking for discrepancies.

Returning to what to include in surveys, consider using scales like importance, satisfaction, and extent of actual behavior. This was done in a study conducted across 15 universities in Ohio for a minority-student retention project in science, technology, engineering, and mathematics (STEM) as profiled in Table 2.3. Students were asked about the importance of retention services for academic work and satisfaction with and frequency of use of same. Some of the content was further probed in a focus group interview. Don't lock into one method; instead, cross-pollinate. In most surveys, questions are usually clustered into labeled groups of related items, and those clusters could become the emphases of an FGI (focus group interview) or an individual interview. This underscores the interrelatedness of different data collection methods.

❖ PRINCIPLE 2: INCLUDE LEVELS 1, 2, AND 3,
  IF POSSIBLE (ALSO SEE PRINCIPLE 4)

What individuals or groups should be sampled for the survey? Would they agree or have quite varied views of a topic, and can radically different perspectives be reconciled? How might such a result affect the assessment, and could it have a negative impact on Phase III (translating needs into solution strategies)? Wouldn't it be better to learn of this before prioritizing needs and recommending solutions? Obviously the inclusion of multiple levels is important. Consider the following:

- If the levels see how their participation had an impact on the eventual actions taken by the organization, ownership is enhanced. (Good communication and publicly crediting groups and individuals for their help are important in needs assessment.)

- Costs and the time for analysis and interpretation rise as data gathering is expanded (but in one study in Table 2.3, a single instrument was used for all constituencies, and via demographic data, they were compared at limited expense).

- When additional groups and constituencies are assessed, *within-method* variations may be necessary; thus more time is needed for item writing, and the needs assessment is more complex.

Including Levels 1 (receivers of services), 2 (deliverers of services), and 3 (the overall system) should be examined in-depth before making final decisions to do so. All the needs assessments in the prior chapter used representatives from multiple constituencies.

Should the responses of different levels be weighted equally? In one study in Chapter 2, responses from Level 3 were viewed as being of greater importance than those from Level 2. A solid rationale was made for doing that, but it may not hold for other settings. If needs assessment is thought of as a democratic process inside of an open and challenging work environment, serious consequences might occur in terms of morale, disenfranchisement, freedom of expression, loss of employees, and so on. Even a seemingly simple choice of what groups to sample could have major implications for the assessment.

Level 1—the direct recipients of services and goods delivered by organizations (businesses, schools, health departments, agencies, etc.)—is automatically a part of the needs assessment. It is the organizational *raison d'être*. Students, parents, clients, patients, and consumers are in the best position to provide perceptions about all aspects of what is delivered. Levels 2 and 3 are there to help Level 1. If an assessment is done only for Level 2, which may be the focus of a training needs assessment, the service deliverers could be thought of as a pseudo Level 1 group. This assumes that the needs of Level 1 are already understood.

Level 2—teachers, health care providers, social workers, and persons employed in businesses—has valuable insights about those being served and what is provided. But there can be problems with these data. In a training context, Holton, Bates, and Naquin (2000) cautioned that what Level 2 individuals say is needed might be *wants* rather than realistic *needs*. This is especially true for data from self-reports of behavior on surveys. Instead of getting at true needs (most needs are relative in nature), we might see what respondents want to get out of the assessment or what they think their supervisors want to hear. "Let's get what we want as opposed to what is really needed!" This happens but can be partly eliminated by the careful wording of questions.

Furthermore, because organizations conduct needs assessments and usually assign the process to Level 2 personnel, there may be a subtle shift from Level 1 to Level 2 concerns. If the NAC strays in this way, the facilitator must get the group back on track. With that in mind, an outsider might review what is being done, looking for such shifting (not having Level 1 as the prime issue).

Level 3—higher administrators, auspice providers such as legislators, and governmental agencies or other groups—appears less often in Phase II. It may be better to have this level come into the picture toward the end of the phase or in Phase III when more is known about needs and actions being considered. This varies with each needs assessment. Another reason for cautiously excluding Level 3 is that the assessment might appear top-down (controlled) even when it is not. If decisions appear to have already been made, it is difficult to get honest responses and perceptions from respondents. "Why put in your 2 cents when you sense it is a lost cause and everything has been decided beforehand?"

Lastly, in fields such as education and in community assessments and capacity-building efforts, think about external groups (concerned citizens, businesses with a stake in the community, senior citizens who do not have children in schools but who pay property taxes that support education). Society as a whole benefits from education—everyone has views on it and knows that it has an impact on many facets of life. When feasible, include broad audiences but take into account their contexts. Many individuals have not been inside of a school for a long time, so it may be wise to provide short descriptions in surveys about what schools currently do and have available for students and teachers.

Remember that obtaining the opinions of more groups and levels increases the cost and intricacy of data collection and analysis. The demands of collating, interpreting, and dealing with divergent viewpoints; finding coherent and communicative means for presenting data; and creating *within-method* variations escalate. Introducing more things into the needs assessment is fine but not necessarily easy and problem-free.

❖ PRINCIPLE 3: EMPLOY MULTIPLE METHODS

Needs assessments are not the province of a single-method mindset. Needs may be subtle with features and dimensions hidden beneath the surface. They must be examined from different angles; are there parts of the problem that should be explored in greater depth than a survey may allow? (This is to suggest not that surveys produce superficial information but that other data are useful for understanding and making decisions.) Or if one method is of a qualitative bent, consider how to combine quantitative procedures into the mix.

Combinations occur in epidemiology where surveys might be utilized with the analysis of records and databases to identify the perceptions of key decision makers. How serious is the problem now or in the future? What policies might have to change as a result of the epidemiological findings? What sort of information campaign should be employed to get information out about an impending problem? Another example of multiple methods is the use of group processes (small discussion groups, FGIs), individual interviews, and observations to round out survey data. Replicate methods to ensure reliability, budget permitting. Two or three FGIs that yield similar results are more persuasive than one. If the NAC is of moderate size and participates in the interviews, do multiple interviews. When possible, allot some funds for replication purposes.

In multiple methods, one strategy might inform the development of another. Individual interviews and FGIs uncover emotions and thoughts about a topic and terms people use in referring to it. (This has been done with panels similar to seated juries to see reactions to different versions of oral arguments in trials.) The information generated is input for survey design and question wording. FGIs may be done after a survey to see how groups interpret and ascribe meaning to results. There are many fun ways to do these types of things.

❖ PRINCIPLE 4: VARY INSTRUMENTATION TO
FIT THE SUBTLE DIMENSIONS OF MULTIPLE GROUPS
(WITHIN-METHOD VARIATION)

The same question order and wording will not always work for different groups—the concept of *within-method variation*. Think of supervisors and employees in a company or of a needs assessment in a school system for the upper elementary and middle school grades where teachers, students, and administrators are surveyed about needs in science and mathematics education. Identical wording and order might not work. Will the same questions be applicable to fifth graders and students in the eighth grade or to employees and supervisors? Surveys have to fit the vernacular and thought processes of the groups being studied. This should improve response and item completion rates but requires more work of the NAC and its facilitator.

In Table 2.3, three of the entries had versions of a *within-method variation* as in the school dropout and violence prevention study where Hunt et al. (2001) described different versions of interview questions for students and teachers. Although done for interviews rather than surveys, the principle is clear. In the evaluation of the STEM program, Altschuld, White, and Lee (2006) examined retention services for minority students in universities. Students accessing or familiar with services and faculty/administrators of retention programs who knew about them through personal experience or interactions with students were the focus of the investigation. This information was sought from the two constituencies via altered versions of the questions. If that had not been done, the questions would not have had as much meaning for Level 2 (Table 3.2).

Table 3.2   Wording Differences Between the Two Surveys—A *Within-Method Variation*

	Importance	*Satisfaction*	*Frequency of Use*
Faculty Survey	Extent to which the service is important to the academic success of students	Your satisfaction with the service for students	Frequency of students' use of this service
Student Survey	Extent to which the service is important to your academic success	Extent to which you are satisfied with the service	How frequently do you use this service?

Source: Adapted from "Effects of the Participation of Multiple Stakeholders in Identifying and Interpreting Perceived Needs," by Y.-F. Lee, J. W. Altschuld, and J. L. White, 2007a, *Evaluation and Program Planning, 30,* p. 3. Adapted with permission.

The *first* feature of the wording to note is that for students it is based on direct and intimate exposure to the service and knowledge and understanding deriving from that experience. *Second,* these questions are asked for personal perspectives, not those of a general group of students. Questions like these may be worded in two ways depending on the purpose of the assessment (personal perceptions or group-oriented ones).

Third, the structure of the questions for importance and satisfaction is intentionally similar. By similar wording and syntax, the subtraction of scores for importance and satisfaction to create a discrepancy score is reasonable. Compare that to more disparate variables such as importance and achievement, although such practice is relatively common in and not questioned in needs assessment. *Fourth,* the surveys have a not-applicable choice and a midpoint (neutral response) on the scale. On scales, "not applicable" is denoted by *NA,* which is not to be confused with the abbreviation as sometimes used to stand for *needs assessment.*

Some surveyors suggest that respondents should make a positive or negative selection and not have a noncommittal response. This may not be sensible in needs assessment where we are trying to learn what a group thinks about a topic and using what is learned for decision making. Mandating artificial choices could lead to erroneous conclusions. Why should a response be forced if respondents are really undecided? The recommendation is to have a neutral point, but others may not agree. The concern is to get an honest view of where people stand on an issue instead of one that is arbitrary. If they have no opinion or they don't know or are unaware, let them state that. If a lot of respondents don't know about current status or don't have an opinion, then the NAC must take that into consideration. If there are highly different rates of these responses across different groups, it reveals a lot about the context. Respondents might not have been exposed to a program or service, they might have no impressions of it, or communication might not be very good. Imbed these rates into needs assessment reports.

One caution is in order about NA (not applicable) and DK (don't know) responses. For double- or triple-scaled items, they lead to different numbers of respondents for each scale. Generally, there are more respondents for importance and less for the other scales. How do you calculate discrepancies when the numbers for each scale differ? Using averages for an item, one scale would have one sample size whereas the other would be based on a smaller number of respondents. Should the discrepancy be just for those individuals who responded to both scales?

This would significantly reduce sample size, and only a subset would be determining the discrepancy score, not the entire group. Such a reduction can be noticeable and cause difficulty in interpreting results. This was observed in the minority retention study (Table 3.3). There were notable differences in NA responses within and across the two groups in the table. The faculty rates ranged from 0% to nearly 30%, and the students went from slightly over 5% to a very high 68% for the importance and satisfaction scales. The issue is apparent.

A statistician might impute (estimate) a substitute score from each group of respondents to get equal numbers for scales. In needs assessment this may distort results and is not seen as commensurate with what the data are telling the NAC. Perhaps a more appropriate way would be to report and explain the NA responses in reports as indicating where a respondent group is in terms of knowledge. Then determine discrepancy or gap scores from a subtraction of averages derived from different numbers of respondents. This is not an ideal solution, but it does not introduce artificiality into the results. (See Lee, Altschuld, & White, 2007a, 2007b.)

Fifth, returning to Table 3.2, look at the faculty version of questions, which takes into account that faculty members have not participated in any service and their views would be vicarious for student services. This underscores the fact that attention should be paid to question wording for different groups. Faculty members are an indispensable part of the culture and environment of the university and science disciplines. They are aware of many of the services, have had students or

Table 3.3 Frequencies and Percentages of NA Ratings for Services From Student and Faculty Surveys

	Student Survey		Faculty Survey	
Category/Variable	# of NA[a]	%[b]	# of NA	%
Precollege Services				
Recruitment to college	26	15.5	0	0
Campus orientation	11	6.6	0	0
Summer bridge programs	63	37.5	3	7.7
Academic Services				
Peer study groups	21	12.5	4	10.3
Tutoring	20	11.9	1	2.6
Supplemental instruction by student facilitators	32	19.1	3	7.7
Coenrollment in courses	61	36.3	6	15.4
Collaborative learning	43	25.6	4	10.3
Living learning program	71	42.3	7	18.0
Drop-in/study center	48	28.6	4	10.3
Glenn-Stokes Summer Research Internship	106	63.1	9	23.1
Summer Research Internship	88	52.4	3	7.7
Glenn-Stokes Academic Year Research Internship	113	68.3	11	28.2
Academic Year Research Internship	105	62.5	4	10.3
Faculty mentoring	55	32.7	0	0
Peer mentoring	58	34.5	2	5.1
Grad student mentoring	84	50.0	6	15.4
Industry representative mentoring	91	54.2	6	15.4

Table 3.3 (Continued)

	Student Survey		Faculty Survey	
Category/Variable	*# of NA[a]*	*%[b]*	*# of NA*	*%*
Financial Support				
Financial aid, grants, and loans	12	7.1	0	0
Scholarships	9	5.4	0	0
Work study programs	44	26.2	0	0
Internships	31	18.5	0	0
Assistance with on-campus employment	37	22.0	3	7.7
Assistance in locating off-campus employment	49	29.2	3	7.7

Source: From *Effects of Multiple Group Involvement on Identifying and Interpreting Perceived Needs,* by Y.-F. Lee, 2005, unpublished dissertation, The Ohio State University–Columbus, p. 92. Reprinted with permission.

[a]Number of NA ratings on each service.

[b]Based on total number of respondents (168 in the student group and 39 in the faculty group).

others talk to them about what is going on, and have opinions and impressions about services. They are key Level 2 respondents, and information from them is highly utilitarian for the needs assessment. The faculty items in Table 3.2 depict a *within-method variation.* It is important to consider where a respondent or his or her group might be coming from, what his or her typical involvement is, and how such factors influence the content/structure of questions. Needs assessments usually benefit from doing so.

The queries used in the study worked well, and students and faculty had relatively close perceptions for most areas in the survey. Items with differences and where the groups were knowledgeable about a service were selected for follow-up with a sample of members of the groups. If there are many items like this, only a subset can be looked at in this manner. In Figure 3.1, the structure of a question from that second,

Figure 3.1   An Example of a Chart Used in the Follow-Up Survey

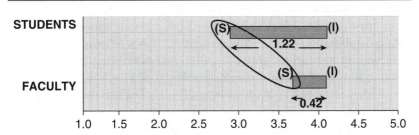

Source: From "Effects of the Participation of Multiple Stakeholders in Identifying and Interpreting Perceived Needs," by Y.-F. Lee, J. W. Altschuld, and J. L. White, 2007a, *Evaluation and Program Planning, 30*(1), p. 3.

Note: The bars and circle in original survey were marked by different colors.

open-ended survey is shown. The loop in the figure was to focus respondent attention on group differences. Simple bar charts may have worked as well. The results indicated that the follow-up was worth the effort, but more costs were incurred. When groups differ, examining why is encouraged.

The open-ended data pointed to some varied understandings about the campus environment. Students did not particularly value or see the need for faculty mentoring, but faculty saw it as important for socialization into a field and career development. This may be expected because students have been in the sciences for a shorter time and may not in their psychological and occupational development have the whole context in view. The follow-up revealed a need for better counseling, guidance, and communication and was useful for thinking about needs.

Locating part-time employment was another area where students felt they were not receiving much help. They observed that faculty members were employed and less empathetic with one student, suggesting that faculty had been employed for many years and could not relate to or have understanding of where the students were. Students cited the poor quality of assistance for finding part-time employment. The faculty members were unaware of this, and the students may have been right about faculty perception of the topic. On the other hand, were students' expectations for the service unrealistically high? The complexity of survey responses is underscored.

The open-ended responses revealed different "world" perspectives that were helpful for explaining how some aspects of the environment might be improved. Without the follow-up, it would have been difficult to fully understand the needs-related data; although it was more work for the NAC, costly, and time-consuming, it was also an interesting combination of two ways to conduct needs assessment surveys. One was quantitative, and the other one was open-ended, based upon initial quantitative results. Sometimes a follow-up is necessary. Other alternatives are to conduct individual and/or focus group interviews.

If many differences across constituencies are uncovered, how is the assessment to proceed? How could differences be reconciled? Are more resources needed to explore the lack of consensus? How many differences can be studied and at what expense? Simple needs assessment surveys may not be that simple, and the word *pesky* in the chapter title was purposeful.

Within-method variations and using multiple groups from the levels may take on varied shapes depending on the creativity of the needs assessors. Holton and colleagues (2000) had different questions for high-level administrators and direct-service providers about performance in a training needs assessment in a large state agency. Administrators may have a broader perspective of what might be needed to improve rather than a parochial one.

Service providers see needs, but there might be more of a wants flavor (see previous discussion) to their responses. Although an expansive view may reside at higher levels of an organization, care is again advised when limiting the input of some constituencies; the matter is delicate for it may reinforce a top-down perspective. If higher-ups are to be given differential consideration and if the facilitator of a needs assessment is external, he or she should tactfully raise some of the concerns that could occur as a result of this stance.

❖ PRINCIPLE 5: CONSIDER USING TWO OR MORE SCALES FOR ITEMS ON THE NEEDS ASSESSMENT SURVEY

Need is the measurable discrepancy between the what-is and what-should-be states, and the survey is a ready-made vehicle for double or even triple scaling. With more than one scale, the amount of information obtained is magnified, and it is easy to develop and format such scales, especially after the importance or what-should-be scale is created.

What is sometimes observed when single scales are used is that there are two patterns of wording; the first deals with the importance of items, and the second is when the word *need* is in a statement—that is, when a person or group needs this or that kind of a program. See Versions 1 and 2 below. Hamann (1997) referred to both patterns as preference scales, not needs-oriented ones. *Need* as a verb implies a solution; *need* as a noun stands for a discrepancy or gap. The distinction is important, and using the word *need* in items is to be discouraged.

Version 1

Rate the importance to you (or the organization) for the statements (what should be) included on this instrument. Use the scale where 1 = very low importance, 2 = low importance, 3 = average importance, 4 = high importance, and 5 = very high importance.

Version 2

Listed below are a number of needs—for each, indicate its importance as a need for you or this organization (*really two different versions of the question depending on whether an individual or organizational perspective is called for*) by using the five rating points where 1 = the lowest value and 5 = the highest. Respondents rate statements like these via five rating points in matrix form to the right of them.

These types of items are observed, but it would be better to extend one scale to two or three to collect more data about needs. If resources are directed toward their resolution, then why settle for less rather than more data? It's worth the extra effort to use more scales. In some entries in Table 2.3, discrepancies were not measured. This was compensated for by other information that was incorporated into the assessments. In every case, a deeply probing understanding of needs emerged from the process.

Going a little further, Table 3.4 is an example of the use of three scales in a needs assessment. The questions and scales are for students with a within-methods version for faculty. When multiple scales are employed, the data multiply per item. For 50 or 60 items of this type, there could be up to 180 data points. The instrument would not be too long, and the yield would indeed be very large.

Table 3.4   Three-Scale Version of the Student Instrument Used in the Minority Retention Study

	Extent to which the service is *important* to your academic success.		Extent to which you are *satisfied* with the service.	How *frequently* do you use this service?
	Strongly Disagree ⇔ Strongly Agree		**Not Satisfied ⇔ Very Satisfied**	**Very Never ⇔ Frequently**
	NA 1 2 3 4 5		1 2 3 4 5	1 2 3 4 5
Financial aid, grants, and loans	O O O O O O		O O O O O	O O O O O
Scholarships	O O O O O O		O O O O O	O O O O O
Work study programs	O O O O O O		O O O O O	O O O O O
Internships	O O O O O O		O O O O O	O O O O O
Assistance with on-campus employment	O O O O O O		O O O O O	O O O O O
Assistance in locating off-campus employment	O O O O O O		O O O O O	O O O O O

Section III

Universities also provide other services as shown below. Rate them in terms of *importance, satisfaction,* and *frequency of use.* If you are not familiar with a service or your campus doesn't have it, denote *not applicable* under the importance column and move to the next item.

Source: Adapted from "Problems in Needs Assessment Data: Discrepancy Analysis," by Y.-F. Lee, J. W. Altschuld, and J. L. White, 2007b, *Evaluation and Program Planning, 30*(3), pp. 258–266.

Note: The circles in the table denote tabs or buttons that respondents selected when responding to the Web-based instrument.

There is another issue about scales that has not been covered. Many options are possible, including the following:

- Likert-type five-point scales using "strongly agree" to "strongly disagree" anchors;

- Likert-type scales ranging from 1 = lowest importance to 5 = highest importance;
- the above scales with values that are gradations such as none, 1%–20%, 21%–40%, and so on and/or similar numerical scales;
- a whole host of other five-point scales with different points on them;
- versions of the scales that use NA, DK, or a neutral or middle type of value and others that do not allow for neutral choices; and
- semantic differential approaches where the respondent chooses from a continuum that has polar-opposite anchors such as warm and cold.

A further alternative is a behaviorally anchored rating (BAR) scale. In work with children, Witkin and others (1979) devised such a scale for needs related to reading practices. A sample of what they did has gently been modified by the author. Students were asked to rate each item on what they could personally do and what students in their grade should be doing rated on an A–E scale (A–E are like grades, demonstrating the use of student terms and what students were used to seeing). Two responses were required for each item. There were unique behavioral anchors depending on the content of the question. The idea was to think about what children would be doing when visiting the library and what they would observe other children doing.

A Sample Item (Adapted Version)

In the library, I find books ... (Students would choose a letter response.)

A	B	C	D	E
Only with someone's help	Between A and C	By using catalogs and reference guides	Between C and E	By using catalogs, reference guides, and the computer

In the library, students in my grade should find ... (Students would choose another letter response for this second question.)

It takes more time to write items with behaviors imbedded in them. But it is an interesting way to construct questions, and the scales are meaningful because of the anchor points.

Many choices are possible for item and survey design. Consider the educational and experience levels of participating samples and choose formats accordingly. If you have the time, pilot test a couple of them. Challenge the group developing the survey. Could the BAR concept be adopted for adults? Look at the fun example of the follow-up survey shown in Figure 3.1; would it fit another needs assessment? What guidance does the literature provide for our survey? Would some formats work better in certain places, and would others be better in a different section of the survey? There are endless questions for the NAC as it produces the survey. Do not despair; ask a few of them and trust in good judgment!

If the audience has limited language ability, use simpler scales (yes, no, uncertain). The loss of information is compensated for by obtaining data from those who are less language proficient and might not answer more complex surveys. Lastly, keep items short and to the point and avoid jargon in the wording of questions.

❖ PRINCIPLE 6: CONSIDER PROCEDURES TO COLLECT INFORMATION ABOUT BARRIERS TO SOLUTIONS AND PREFERRED WAYS TO IMPROVE

Even though the main emphasis of Phase II is information about needs, the survey could have questions about barriers in the organization or how solutions could be implemented. These were prominent in one of the studies in the previous chapter.

In needs assessments conducted for training purposes, sometimes preferences for how training might be delivered, costs associated with it, the best times (days, months, seasons) for implementation, time available for employees to participate, where the training might take place, and motivation for learning are included. Use questions like these, especially if there is only one opportunity to reach the sample. Place them in the latter portion of the instrument, not earlier, except where the main intent is to uncover barriers to solutions or explore options regarding them. A few examples are in Table 3.5. They are helpful for thinking about solutions, so take advantage of the opportunity with perhaps 20% or so of the survey devoted to such issues.

About barriers or impediments to dealing with needs, some agencies, companies, and institutions may not truly support (funds, release time, encouragement) staff training. The demeanor is "Yes, we give lip service (tacit support for the activity), but in reality we don't really want you to do much of it, or it takes time away from productive work." In essence there is a disparity between what organizations say and what they do.

If in Phase I the NAC gets this sense of the environment, probe into it. Ask about the receptiveness of the organization to change, factors that might reduce the likelihood of improvements being successful, the nature of backing for new directions, and other similar topics. Serious

Table 3.5   Examples of Questions Regarding Purposes, Modes of Delivery, and Barriers in Training Needs Assessment Surveys

General Focus	Sample Ideas for Questions
How training might relate to one's work/career	Enhance job performance
	Lead to more job satisfaction
	Improve job security
	Give one a competitive advantage for promotion
	Credits (continuing education and the like)
	Other reasons for training supplied by respondent
Preferred modes of training	Traditional classes
	What locations would or would not work
	Self-study (distance) with practicum
	Distance program
	Time involved, preferred time periods for taking the training
	Most favorable times of year
Barriers to training	Level of employer support (dollars, time)
	Level of employer support (enthusiasm, encouragement)
	Personal cost factors
	Inability to use training on the job
	Motivation
	Family factors

problems might be detected this way. Be sensitive and careful in writing statements about how the organization functions, and the issue should be discussed with the NAC before decisions are made. The questions could be threatening; nevertheless, the survey is about needs and is a good way to gain such information.

There are many examples of surveys and needs assessments in many fields that focus primarily on the nature and delivery of training. Among other things, their questions deal with barriers, support for the activity in terms of value and budget, likelihood of participation, areas including maintenance of strengths, and relation to personal and professional development. A few references are provided to such assessments (ABLE Design and Evaluation Project, 2008; Conklin, Hook, Kelbaugh, & Nieto, 2002; University of North Carolina Center of Excellence for Training and Research Translation, 2006; Wilkie & Strouse, 2003). While needs are a concern of these studies and others like them, the general intent is more of a combination of the assessment of need, the feasibility of potential solutions, and an examination of contextual factors.

❖ PRINCIPLE 7: CREATING THE NEEDS
     ASSESSMENT SURVEY AND OPEN-ENDED QUESTIONS

Examples of open-ended issues that require careful thought from respondents include the following:

- giving illustrations of problems they are seeing that need attention;

- describing potential solution strategies for gaps;

- indicating if the survey was a meaningful way to capture information about problems;

- describing barriers to improvement; and

- providing other thoughts and ideas about conditions and what it might take to improve them.

Use such questions with one proviso: Even with programs for analyzing qualitative data, interpretation can be tricky, and experience in connecting thoughts and concepts in responses is needed. Most facilitators have a background in doing so, and hopefully some members of the NAC will as well. What are the key variables in the data? What are the main themes in the data and overarching ones that link the main themes? What is the explanatory power of the latter for understanding of needs? Have several members of the NAC examine and analyze open-ended data independently. Do the independent summaries agree?

Well-framed open-ended questions can lead to valuable responses. If we ask people "what if" or "what might another person see" if he or she looked at an issue or to describe how the organization might respond to a specific circumstance, replies may be voluminous and rich. Usually, probes are included with items to prompt in-depth consideration of ideas. Think about including between two and four such questions. Strive for a balance of mostly scaled items and open-ended ones. Balance makes the survey a workable proposition. Remember that open-ended questions do not directly lead to discrepancies, so integration of results with those from other sources is necessary.

The seven principles with the examples in the text provide guidelines for an NAC as it begins the survey process. In Table 3.6 we return to them in more of a checklist way. They are helpful, but note surveys are in reality partly art.

Table 3.6   The Needs Assessment Survey Checklist by Principle (Translating Principles Into Action Steps)

Principle	Steps
Decide upon content	Review prior work (Phase I) and reports for ideas.
	Consult the literature for theory and for survey work in other needs assessments like yours.
	Make decisions as to what is relevant to your local needs assessment.
	If there is too much content to cover, consider prioritizing it or if alternative versions of surveys might be developed.
Inclusion of samples from Levels 1, 2, and 3	Carefully identify the samples to include from Levels 1, 2, and possibly 3.
	Are there subgroups within levels that should be in the needs assessment such as younger and older students, upper and middle management staff, and so on?
	Select levels in accord with the purpose of the needs assessment, budget, time, and human resources available for the job of developing and implementing instruments.
Employ multiple methods	Are there areas where the survey process might be augmented by other methods?
	Might it be best to conduct individual interviews, FGIs, or observations prior to designing/implementing the survey?
	If the survey may not yield the depth of information desired, should it be followed up to provide more understanding of needs?
	Consider, before undertaking multiple methods, how the data could be integrated into a holistic picture of needs.
Multiple versions of surveys (within-method variation)	Look at the samples for the survey process to see if subtle versions of wording and orders may be necessary.
	The NAC should put itself in the shoes of respondents to see if the situation requires different versions.
	Pilot test the versions, if time permits, to see if the wording is appropriate.

(Continued)

Table 3.6 (Continued)

Principle	Steps
Use multiple scales	See examples in the earlier text.
	In general, use response categories with NA, DK, and neutral points on the scale.
	To create discrepancy scores, have wording in item stems that is fairly similar to enhance the rationale for subtraction.
	As in the prior principle, think about respondents' comfort zones in responding.
	Have two or more scales to increase the information obtained.
Include questions about barriers to solutions and/or preferences especially for training needs assessments	In training needs assessments, ask about preferred modes of training, impediments/barriers to training, support for the endeavor, and so forth.
	Word such questions in a sensitive, nonthreatening manner.
	Place such questions near the end of the survey so as to not confuse purposes.
	It may be possible to have some questions on the survey about solution strategies, but this is not the most common of needs assessment procedures.
Assemble the final survey and plan for distribution	Make a draft of the survey to get a sense of what respondents might see.
	The NAC should, individually and as a group, critically review the draft with an eye toward improving it.
	Think about the best means to deliver the survey to potential respondents.
	- If respondents are geographically dispersed, computer literate, and connected to the Web, look into Web-based surveys (many commercially available survey mechanisms are fairly inexpensive).
	- If respondents are in naturally occurring groups such as schools, companies, agencies, and/or conferences, think of distributing the survey at a group meeting where in all probability it will be completed at the time of distribution.

Highlights of the Chapter

Given that Table 3.6 is like a checklist, the highlights are brief.

1. The chapter was intended as a framework for surveys in needs assessment.

2. Hopefully the principles will help needs assessors to produce quality surveys.

3. The survey is not usually the sole needs assessment method. Coupling data together enriches understanding and makes for a better assessment.

4. Use the examples of scales, formats, and questions in the text to your best advantage.

5. In Chapter 4, there is an introduction to epidemiology as applied to the needs assessment context. It, along with surveys, is one of the major quantitative methods employed in needs assessment.

4

Basic Epidemiology
for Phase II

❖ SOME BACKGROUND

Epidemiology is a way to understand the current situation and what might take place in the future primarily from information in major data sets. It comes from the study of diseases, and its purposes are to determine the prevalence (how many cases of the disease currently exist) and the incidence (how many cases might occur over time), identify causes and eventually the most probable cause of the disease, learn about current treatment and the effectiveness of same, and link results to policies and actions to relieve needs that have been uncovered. If linkage is not established, the epidemiological study may be technically valid but not successful practically. It doesn't help to see a rapid rise in a disease (diabetes in children and adults) without reducing the rate of occurrence and/or curing the disease.

The second purpose illustrates the complexity and adventure of the approach. Consider such outbreaks as avian flu in China (Fairclough, 2009), salmonella in Ohio (Crane, 2009), or a local case of food poisoning. Tracking down cause involves detailed detective work as to the most likely causal factor for the malady. It is a painstaking undertaking based on collecting a variety of information and continually sifting through a great deal of data and findings to pinpoint the most probable culprit. For food poisoning this would entail looking at what foods were eaten, where they were consumed, and, if people ate the same foods, which among them might be the one leading to the outbreak.

The latter example is from Hebel and McCarter's (2006) discussion of an outbreak of food poisoning in schools and children in a rural area. Questions that might be asked are:

- Do we know if we have evidence of an epidemic (some people routinely get sick to their stomachs, so do the observed cases exceed normal expectations)?
- What other information should be sought to explain what is happening?
- Is it possible to pinpoint when the initial event started?
- Is the problem more apparent in certain age groups of students and/or school populations, and what possible (causal) factors do those groups have in common?
- What conclusions can be drawn from the above steps (rates are more than expected, certain groups are more affected, potential causes seem to be by location of exposure, or a combination of factors)?

While much of epidemiology comes from database analysis, it also has other components. Qualitative techniques, interviews, and observations are often employed. Cohen (2006), for example, used intensive interviews of knowledgeable individuals (key informants) as part of an epidemiological study of public health capacity in Canada. Similarly, Dance, Brown, Bammer, and Sibthorpe (2004) conducted a culturally sensitive survey as part of a public health investigation into current and future needs of an aged native population.

The purpose of this chapter is to explain the use of epidemiology and related methods (futures forecasting) for needs in social services, rehabilitation and corrections, mental health, education, insurance, finances, and governmental services besides their use in health. These other fields are awash with databases ripe for epidemiological analysis.

❖ THINKING ABOUT DATABASES: AN OVERVIEW

From databases it is possible to see what has happened in the past, what is happening in the current situation, and what may take place in the future. Predictions that go further and further out in time, however, are more tenuous due to other events and variables entering the picture. These may be new treatments for diseases or public awareness campaigns that affect what is observed. The use of cigarettes as an illustration has seen increases in antismoking campaigns and changes in municipal ordinances prohibiting smoking in certain locations.

One good database source is the national census. Data on many variables (housing, age, race, etc.) are routinely entered and maintained in a storehouse of information that is mined by those who establish public policy or conduct needs assessments for companies, educational and government agencies, and other organizations. Many large organizations (public and private) have similar capabilities. Their bases are vast and have been in place for extended periods. Setting up new entities like these or modifying existing ones is expensive and not usually done by small agencies, institutions, or businesses due to budget constraints. Rather they seek what is available from larger systems as related to needs and topics of concern to them. Hopefully the data will be suitable for providing direction and guidance.

An example of a major database will demonstrate how immense and valuable information is and the cost involved in obtaining and maintaining the archival trove. The U.S. government has been amassing information on a large number of variables about the health status of American youth. The National Health and Nutrition Examination Survey (NHANES) is an effort to monitor the well-being of children and to alert the country as to what might be necessary for maintaining current status when appropriate or improving it, if indicated. For NHANES refer to the citations for the Centers for Disease Control and Prevention (1999–2002) in the reference list. Also, the World Health Organization has myriads of health data. These are particularly important for when avian or pandemic flu arises. Disease does not respect borders. The same kind of thinking applies to animal populations such as deer; the chronic wasting disease of these animals cuts across countries with decimating effects on herds (Kumar & Altschuld, 2004).

Returning to American children, toward the end of the 20th century, the government decided to enhance what it was doing to collect health data. The variables in NHANES were expanded. Carefully selected individuals between the ages of 5 and 19 were brought into specially designed mobile assessment centers for detailed health evaluations. Since most subjects were not of legal age, appropriate approval by parents or guardians was required.

Young individuals come to transportable centers that are set up in locations throughout the country. A center is a group of semi-tractor-trailers linked together to form a complete diagnostic facility with a full complement of trained health professionals. Data are collected, and an in-depth picture of the health of the nation's youth is determined. Figure 4.1 is a schematic for such a center. Due to costs this almost certainly has to be done by the federal government.

At a site, individuals receive physical examinations consisting of tests related to well-being (blood, measurements of height and weight, body-mass index calculations, blood sugars, etc.). Surveys are used to obtain data about lifestyles, amount and types of activity undertaken daily, what and how much food is eaten per day, and so on. Based on the sampling procedures, inferences to the total population of youth in the United States are made.

Sizable funds are needed for a system of this scope and continuing it into the foreseeable future. NHANES is about the health of a sizable part of the population of a large country with many implications for policy and costs that could be incurred over time. It is critical for making wise decisions and establishing national priorities. Think of the risks to youth and what could happen without implementing and continuing the database. Preventing later problems is worth the up-front expenditures. Would we know where to put resources without information coming from NHANES?

Everything in the system and what is learned from it is available free of charge to the general public, researchers, and interested parties in the same manner as the census. Because the original and modified NHANES systems have been there for a long time, it is possible to look at the current status and emergent trends that should be attended to before problems get too great. (Caution in calculating trends must be

Figure 4.1   Schematic of the Mobile Diagnostic Center

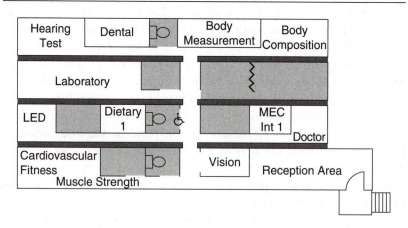

Source: From *Examination of the Determinants of Overweight and Diabetes Mellitus in U.S. Children,* by J. M. Chiasera, 2005, unpublished doctoral dissertation, The Ohio State University–Columbus.

exercised as explained subsequently.) It may also be possible to tease out causal factors through sophisticated statistical techniques.

In 2005, Chiasera looked at childhood diabetes and the onset of adult diabetes in children and what the data set might tell us about the overall well-being of this group. What might be occurring now, and what is likely to appear over time? What might be the impact on the health care system if the disease is left unchecked, and what will be needed to treat it or reduce some of its more severe outcomes? What other associated health problems might occur down the road? What would be the effect on already strapped health care budgets? What kinds of treatments are available, and how effective are they?

Chiasera (2005) found that the percentage of children overweight or at risk for being overweight increased dramatically over 5 years with this indicator being a possible predictor of the disease. This is shown in Figure 4.2. Since the data come from large national samples, they represent an alarming trend. A percentage or two of change contains somewhere between 500,000 and a million individuals. Table 4.1 (Chiasera, Taylor, Wolf, & Altschuld, 2008) offers additional data regarding childhood

Figure 4.2   Proportions of At-Risk for Overweight and Overweight in Children From NHANES 1999–2002

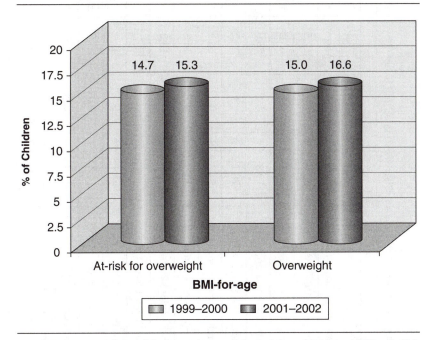

Source: From *Examination of the Determinants of Overweight and Diabetes Mellitus in U.S. Children,* by J. M. Chiasera, 2005, unpublished doctoral dissertation, The Ohio State University–Columbus.

Table 4.1 Mean Differences in Factors Related to Diabetes by Diabetes Status in U.S. Children (Chiasera, 2005)

	Total			Normoglycemic			Pre-Diabetes			Diabetes		
Measures of Adiposity	Mean	±	SE	Mean	±	SE	Mean	±	SE	Mean	±	SE	P
BMI percentile	62.6	±	0.9	62.5	±	0.9[a]	70.7	±	5.4[b]	59.0	±	4.7[a]	0.037
Weight (kg)	63.4	±	0.4	63.2	±	0.4	68.0	±	3.5	65.3	±	3.1	NS
Waist circumference (cm)	79.8	±	0.4	79.6	±	0.4	83.1	±	2.6	82.2	±	2.4	NS
Arm circumference (cm)	28.2	±	0.1	28.2	±	0.1	29.5	±	1.0	29.3	±	0.8	NS
Thigh circumference (cm)	50.2	±	0.2	50.2	±	0.2	52.3	±	1.4	51.0	±	1.4	NS
Triceps skinfold (mm)	15.9	±	0.2	15.8	±	0.2	17.1	±	1.5	16.2	±	1.7	NS
Subscapular skinfold (mm)	13.0	±	0.2	13.0	±	0.2[a]	14.0	±	1.4[ab]	15.1	±	1.1[b]	0.039
Nutritional Status													
Albumin (g/L)	45.4	±	0.09	45.4	±	0.09	45.3	±	0.46	45.5	±	0.52	NS
Red blood cell count (SI)	4.81	±	0.02	4.81	±	0.02[a]	4.87	±	0.09[ab]	4.98	±	0.05[b]	0.003
Hemoglobin (g/dL)	14.2	±	0.06	14.2	±	0.07[a]	14.1	±	0.20[a]	14.9	±	0.22[b]	0.016
Hematocrit (%)	41.9	±	0.17	41.9	±	0.18[a]	41.9	±	0.63[a]	43.8	±	0.58[b]	0.018

Measures of Adiposity	Total			Normoglycemic			Pre-Diabetes			Diabetes			P
	Mean	±	SE	Mean	±	SE	Mean	±	SE	Mean	±	SE	
Dietary Intake													
Energy (kcal)	2,316	±	28	2,321	±	27.00	2,373	±	225	1,970	±	189	NS
% of energy (kcal)—EER (Estimated Energy Requirement)	110.5	±	1.2	110.9	±	1.2	104.9	±	8.3	93.5	±	8.0	NS
Total fat (g)	83.7	±	1.3	84.1	±	1.3	80.8	±	7.1	70.9	±	8.8	NS
Total saturated fatty acids (g)	29.5	±	0.5	29.6	±	0.5	27.5	±	1.9	24.4	±	3.1	NS
Discretionary fat (g)	66.9	±	1.2	67.2	±	1.1	63.8	±	5.7	56.8	±	7.2	NS
Added sugars (tsp)	30.3	±	0.7	30.3	±	0.7	32.5	±	4.7	29.5	±	5.3	NS
Glucose Metabolism													
Glucose (mmol/L)	4.8	±	0.02	4.7	±	0.01[a]	6.0	±	0.05[b]	6.0	±	0.52[b]	<0.001
Glycohemoglobin (%)	5.14	±	0.02	5.11	±	0.01[a]	5.27	±	0.04[b]	6.55	±	0.73[b]	0.001
C-peptide (nmol/L)	0.69	±	0.01	0.68	±	0.01[a]	0.95	±	0.07[b]	0.90	±	0.06[b]	0.000
Insulin (pmol/L)	78	±	3	76	±	3[a]	128	±	12[b]	118	±	14[b]	0.000

(Continued)

Table 4.1 (Continued)

Measures of Adiposity	Total			Normoglycemic			Pre-Diabetes			Diabetes			P
	Mean	±	SE	Mean	±	SE	Mean	±	SE	Mean	±	SE	
Markers of Chronic Disease													
Total cholesterol (mmol/L)	4.12	±	0.02	4.12	±	0.02	4.11	±	0.09	4.26	±	0.21	NS
Triglycerides (mmol/L)	1.03	±	0.02	1.02	±	0.01	1.15	±	0.13	1.05	±	0.10	NS
LDL cholesterol (mmol/L)	2.39	±	0.02	2.39	±	0.02	2.42	±	0.08	2.25	±	0.18	NS
HDL cholesterol (mmol/L)	1.26	±	0.00	1.26	±	0.00	1.16	±	0.04	1.22	±	0.06	NS
Boys	1.21	±	0.01	1.21	±	0.01	1.17	±	0.05	1.10	±	0.05	NS
Girls	1.31	±	0.01	1.32	±	0.01[a]	1.14	±	0.05[b]	1.37	±	0.08[a]	0.009
Homocysteine (umol/L)	5.84	±	0.07	5.83	±	0.07	6.10	±	0.43	5.84	±	0.65	NS
Systolic blood pressure (mmHg)	108	±	0.29	108	±	0.29	111	±	1.82	111	±	2.01	NS
Diastolic blood pressure (mmHg)	63	±	0.47	63	±	0.48	61	±	2.09	64	±	1.90	NS
Albumin, urine (mg/L)	44	±	4.53	45	±	4.68	46	±	17.32	25	±	7.19	NS

Measures of Adiposity	Total			Normoglycemic			Pre-Diabetes			Diabetes			P
	Mean	±	SE	Mean	±	SE	Mean	±	SE	Mean	±	SE	
Blood urea nitrogen (mmol/L)	4.0	±	0.06	4.0	±	0.06	3.9	±	0.19	4.3	±	0.26	NS
Uric acid (umol/L)	303	±	2	302	±	3[a]	328	±	11[b]	344	±	16[b]	0.009
Creatinine (umol/L)	56.3	±	0.96	56.2	±	0.99	54.9	±	2.61	61.6	±	3.07	NS
Inflammatory Response													
White blood cell count (SI)	7.09	±	0.07	7.08	±	0.07	7.48	±	0.39	7.28	±	0.39	NS
Lymphocyte (%)	32.62	±	0.23	32.69	±	0.23[a]	32.23	±	1.52[ab]	29.38	±	1.09[b]	0.023
Segmented neutrophils (%)	55.57	±	0.28	55.50	±	0.29[a]	56.02	±	1.75[ab]	59.04	±	1.56[b]	0.016
C-reactive protein (mg/dL)	0.16	±	0.01	0.16	±	0.01[a]	0.23	±	0.07[ab]	0.24	±	0.04[b]	0.007

Source: From *Examination of the Determinants of Overweight and Diabetes Mellitus in U.S. Children,* by J. M. Chiasera, 2005, unpublished doctoral dissertation, The Ohio State University–Columbus.

diabetes. The variables in the table are indicative of the immense nature of NHANES. It permits an intensive investigation of possible predictors of diabetes. (For other studies see Chiasera, Taylor, Wolf, & Altschuld, 2007a, 2007b, 2008; Taylor, Wolf, & Chiasera, 2006.)

Epidemiological research informs the health community of current status, but with ongoing collection it allows estimates of what is likely to be. Will what was observed by Chiasera and others hold up if the database is looked at 5–10 years from now? Given the amount of data, the key and important variables, the scientific sampling procedures, and the careful way in which the data were obtained, estimates from the base should be quite accurate. Serious situations and how they are changing will be evident especially in terms of needs. Epidemiology is an excellent method for assessing needs.

❖ THINKING ABOUT DATABASES: CONSIDERATIONS

There are some cautions about the methodology.

1. Collecting such health data is very expensive. Local health departments have databases, but the national one is more than most can afford. The funding is prohibitive for initial and continuous data collection and the quality and integrity of the system. Specialized personnel are required. Local bases, while valuable, would not be adequate for work such as Chiasera's.

   (It is important to note two major data collection efforts regarding health, the WOSCOPS and the Framingham study, and what they have revealed about adult health, particularly heart disease and what might be potentially causing it. Through them, the link of smoking and high blood pressure to heart problems was found. It is doubtful that smaller bases would have been up to the task.

   WOSCOPS, the West of Scotland Coronary Prevention Study, and the Framingham endeavor are long-term in nature. The 1951 publication of Dawber, Meadors, and Moore gives an indication of the length of time the latter has been in operation. These databases have spawned hundreds of investigations. It's a relatively direct matter to find studies from the two bases by a search of the Internet and specialized work that would fit the focus of localized needs assessments. In HIV prevention, see Whitmore, Zaidi, & Dean, 2005, for an analysis of the array of data-based information that is available.)

Given costs, the first thing needs assessors must do is to see what data exist and what are their scope, quality, and accessibility. Does what is available align with the concerns/needs of importance to the organization?

2. Inferences from databases are made from the general (the population that provided the data) to the specific local situations, and ecological fallacies can occur (the overall picture might not translate to that of the immediate location).

Witkin and Altschuld (1995) cited work where the nature of a local population was not fully considered when information from a national database was used to calculate rates of asthma in youth. This led to a serious underestimate of how many children in one ethnic group would have the disease, and in turn, inadequate state funds were allocated for treatment.

Being aware of this problem greatly reduces it. It is less probable in NHANES where careful and exacting sampling strategies were carried out along with getting extensive demographic data on participants and subpopulations. Thus it is feasible to disaggregate segments of the population from the overall data. The ecological fallacy is not likely whereas, in other situations, it depends on what is in the base and the specific population that is the focus of the needs assessment.

Always consider the specific groups in need. As a case in point, Garibaldi, Conde-Martel, and O'Toole (2005) found differences depending on the age of the homeless (younger or older than 50). Similarly, Ismail (2004) observed that one implication of epidemiological data for dentistry is that the field must go beyond the "drill" and "fill" mentality for caries to a prevention focus particularly for younger populations. Epidemiology only goes so far before we begin to deal with attitudes and perspectives that affect policy and behavioral change.

3. Sometimes data from bases can be misleading. Goering and Lin (1996) reported that in one Canadian province it was difficult to identify mental health problems due to how data were entered into the system. Physical problems took precedence over mental health ones, masking the mental disorder even when it was the cause of the physical difficulty.

Needs assessors should not hesitate to ask how the data were produced and what rules were followed when entries were made into the base. Are there checks on reliability and quality? If entry is inconsistent, could results be misinterpreted without knowledge of protocols used for transferring the original data? For instance, write-ups on mental health intake forms from the judgment of the person doing an intake interview may vary and go through a retranslation as fields in the database are filled. Might different individuals see different things and not put in the same entries?

Ask about problems or flaws in the database and how they might affect interpretation. Don't be bashful in pushing those who intimately know the nature of how the base was constructed. Are there issues with basic variables? Are certain variables composites of others?

4. The database is valuable to the needs assessment to the extent that it contains information about the needs of concern or interest. For childhood diabetes, Chiasera (2005) observed that the only data on physical activity in NHANES came from self-report answers to survey questions. The validity of these data may be limited, and they do not afford inferences to school physical education (PE) programs or other health-related efforts.

What schools offer in PE was not part of the data. How might well-being be affected with more intense activity programs for children? It is common knowledge that PE programs have been cut or are in jeopardy of further reduction. It would have been useful to know if healthier individuals came from schools with more pronounced efforts and what was contained in them (aerobic conditioning, improving abdominal strength, etc.), the duration and frequency of activities, and their overall quality. Are they geared to deal with the health of children now or in the future?

5. Another issue inherent from Point 4 is the need to connect the information from NHANES to other database(s). This necessitates understanding of relational databases and having access to and using them. Will identifiers in NHANES match up with those in another base to facilitate doing this? This may require specialized expertise beyond the capabilities of the facilitator or the needs assessment committee (NAC).

6. Sometimes even with well-established databases, terms and definitions designed to standardize input shift over time. When examining data over a long time span, be aware of changes in

meaning and be cautious when interpreting analyses. It might not be easy to compare data collected in different periods and draw reasonable conclusions.

(The grandchild of the author has a disease that has been clearly recognized and diagnosed in the last 20 plus years and is treatable. Before that, children died in their teens with little idea of the cause of death; it was just a chance occurrence. Earlier databases were simply inadequate.)

A time-related problem popped up when the federal government expanded group designations for individuals in accord with the evolving ethnic diversity of the U.S. population. Friedman, Cohen, Averbach, and Norton (2000) described bridging (statistical) formulas to compare data from different periods. Care must be taken when applying such adjustments as they are based on varied assumptions. Friedman et al. also noted that a number of federal health-related agencies had distinct requirements for data that were not the same.

Data systems are not static and may exacerbate looking at results from one location to another (states and cities). Are crime rates from one city or state based on the same categories, definitions, and rules of entry? This is apparent in regard to rapes (male assaults on females) where there are many camouflaged aspects as to how the basic data were obtained. The reported rate is probably lower than the actual one due to the trauma and stigma that victims feel, their fear and desire to avoid having their characters assassinated in court, or even having the wrong law enforcement personnel present when the crime is being investigated.

The nature of the data may also be changing. Holding (2007) reported that some states in the United States are considering mandating that the identity of the accuser in a rape case be made public along with that of the accused. This could further reduce reported rapes and make what emerges from database analysis more suspect.

Other variables may have rates lower than the real ones, and the problem or gap in these instances can be less or more than it really is. Think of spousal and/or child abuse, senior abuse, driving while under the influence of drugs or alcohol, family problems, daily caloric intake, and smoking patterns. Some underreporting of prevalence is to be expected. Will individuals be honest about some behaviors, in self-reports and/or interviews, especially when they know they are not socially desirable? Do we conveniently forget that candy bar earlier in the day when replying to a survey on food consumption? Adding to the complexity is when the population surveyed consists of service providers who may overstate (probably unconsciously) the nature of some problems.

An example of a parallel situation is school graduation rates. In some areas they may not be accurate because those who dropped out have been incorrectly counted in the denominator of the fraction. Test results of seniors in these districts appear elevated due to weaker students no longer being there, thus portraying a potentially misleading state of affairs. See Example 4.1.

Example 4.1

From Where Do Numbers Come?

Members of an upper–middle-class suburb were aghast that they had been outperformed by the large, less well-to-do, more ethnically diverse, urban district in the area on a specialized and prestigious statewide testing program in mathematics. How could this have happened?

One possible explanation came from the actual numbers participating. The urban district, although 10 times larger, had about one fifth of the number of students from the suburb taking the test. Perhaps the best students from the former were tested whereas a much broader range of individuals in the suburb took part.

In other words, in one district the students may have been "skimmed" for the test, and in the other all students may have been encouraged to take it. We should not necessarily be skeptical of numbers, but at the same time that should not deter us from inquiring about how data were generated—from whence did they come?

---

7. The last caveat is that causal models developed from epidemiological data are helpful but do not automatically reveal what solution strategies should be considered or would be appropriate. There may be multiple causes of a problem and multiple solution strategies. Each solution will have its own unique character. It was observed in a large state that less-than-desirable numbers of children were receiving appropriate vaccinations. The data indicated the problem but provided little information as to why it was occurring.

This theme is frequently repeated in epidemiology when moving from data-based conclusions to forming and implementing useful policies and procedures for alleviating problems. Numerous authors have cited this issue (see Bugeja & Hwang, 2000; Ismail, 2004; and Knight & Meek, 2003).

Only some fundamentals of the epidemiological study of needs will now be given. Sophisticated epidemiology may involve complex statistical procedures (causal modeling and structural equations) that require special skills. The same idea is true for some of the database manipulation noted before. Many needs assessors may not have the appropriate background and will encounter difficulty if the manipulation is complicated. Additionally, one short chapter in a needs assessment book will not an epidemiologist make! What is done here is to show the value of the method and what is involved in implementing it. The following steps provide a general sense of the process.

❖ THE GENERIC PROCESS OF EPIDEMIOLOGY

Step 1: Make Sure You Need to Do It

The NAC must be clear that epidemiology fits Phase II and understand what it will provide for the endeavor. This holds for any method in Phase II but particularly for in-depth epidemiology (identifying prevalence and incidence, linking relational databases, determining causal pathways, collecting additional information about causes mostly through qualitative fieldwork, gathering data on treatments for problems, and integrating the obtained data). The method, although heavily relying on available data, can still be a dear enterprise, so think about the following:

Do you want to think mostly in terms of prevalence and incidence?

Will there be a necessity to collect causal data beyond what is located in databases?

What would you like to learn about treatments and their effectiveness?

How extensive might these efforts be?

Is the budget sufficient for conducting such work?

What is the level of expertise in the NAC and/or in the organization?

Step 2: Find Out What Currently Exists

2a. Divide the NAC Into Small Groups for Locating Data Sources

Divide the NAC into small groups, with each seeking potential sources available from organizations and agencies.

2b. Searching for Potential Sources

The groups contact local, regional, or state health agencies to identify existing data sets if health is the concern, and the process is analogous for education, social welfare, business (economic forecasts, job conditions, company expansions), alcohol and drug abuse (mental health and substance abuse bases), violence and crime in communities (police and law enforcement groups, data on incarcerated populations), and so forth. Archived information exists, and advantage should be taken of what is there.

Look at the Web sites of state agencies and planning groups to see what they have. Peruse reports that they produce and are accessible to the public. Are trends being observed, what is their direction, and what are the implications? Contact local librarians for help in locating information and what recent reports might be available. Scan what is there to get a feel for its utility for the needs assessment. This is similar to what might have been done in Phase I of needs assessment.

Step 3: Think About the Use of Indirect Estimation

Indirect estimation is where certain variables derived from databases might predict need without any new data being collected. Tweed and Ciarlo (1992) looked at seven different formulas for predicting mental health needs in a geographic area. One of the formulas appeared to be much better for prediction. The SLEM Linear Regression equation, with variables defined, is shown below.

Formula: Predicted Rate = B0 + B1X1 + B2X2

Variables: X1 = percentage of households composed of one person

X2 = percentage of separated or divorced males (age 15 or older)

B0 through B2 are weights used to approximate the population in need of services. After calculating the estimate of those needing help, the needs assessors collect data about the actual numbers receiving help and compare them to the predicted values to see if there are gaps.

Indirect estimation is of value, but there are a few caveats about it for a local needs assessment. First, it is a general indicator that requires drilling down when more detailed aspects (age groups, gender, specific mental health or other problem areas) are of concern. Second, it predicts group, not individual, behavior. It is more difficult to predict what individuals will do and what needs they may have. In rehabilitation and corrections, false negatives and positives have been observed in looking at recidivism for pedophiles. So the admonition would be to consult the literature not only for models like SLEM but also for where problems might occur.

Step 4: Engaging in Conversations With Key People

Armed with knowledge from Step 2, find individuals who are aware of the sources—what they do or do not contain, how easy they are to work with, how they have been utilized, what types of reports have been generated, and how accessible the data are or information is. Seek their input on what the NAC is contemplating.

- How doable is the task?

- How complex might it be?

- If it is complicated, are there any shortcuts or simple ways to extract useful data?

- What costs might be necessary to do this?

- What time frames might it take for the NAC to get what it desires from the base, and how might results be used for decision making?

- Are there reports of which they are aware and that could inform the needs assessment process?

- Are there any problems with data in the area of focus?

- Are there facets of the problem where perhaps not so much is known or available?

- What pitfalls has the NAC observed, and how would it avoid them?

- Are there quick ways to reduce costs without sacrificing much depth and quality?

If help for the study is indicated, ask them as to who might be providers. If the task is not too burdensome, see if they would be willing to assist.

Step 5: The NAC Reconvenes and
Synthesizes What It Wants to Do Next

If the NAC decides it wants to do an epidemiological study, it may also consider multiple methods. A survey could be sent to key decision makers, or perhaps a focus group and/or small discussion group could be used when results are in hand and insight about them is desired. It is easy to combine the numerically rich epidemiology with qualitative and/or survey methods.

Another thing that might be done is to compare data from different (not relational) databases. Passmore (1990) did a study of worker accidents in the mining industry. From a national database, rates for different types of accidents related to the age of miners, their training, and other variables were calculated. Passmore then located data about worker injuries and accidents in other occupations to which he contrasted his findings. The result was useful in demonstrating how dangerous mining is. He also was able to investigate what might be causing accidents by looking at worker training and experience in regard to compliance with safety regulations.

Step 6: Conduct the Epidemiological Analysis

6a. Collect and Array Data as Related to the Area Under Study

Create tables containing what is known about a disease, educational concerns, or other variables of interest. Make sure that they are broken down by population groups related to the focus of the needs assessment. These are basic tables that convey a sense of the issue or problem.

Tables 4.2–4.4 from the study of homeless individuals are examples. In Table 4.4, data about the use of services are presented, which is typical of epidemiology in that the study goes beyond counting how many people have or will have a problem to strategies for resolving it and the extent of their utilization by target groups. Are there gaps in services, and what might be causing them? For the homeless, an interesting finding was that older individuals have more health care benefits (from military service or prior employment) that could be directed toward treatment in hospitals or other facilities. Surprisingly, perhaps because of fear, self-consciousness, or mental health, this older population preferred to use street clinics for help. Knowing this fact gives service providers and needs assessors deeper insight into how to work with this troubled group. Without

Table 4.2   Selected Descriptive Data of Homeless Individuals by Age Group

Variable	50 Years Old or Older	Less Than 50 Years Old
Have Health Insurance	84.7%	56.8%
Have Social Security	41.9	13.8
Income From Friends and Family	8.1	23.9
Money From Hustling and Stealing	5.4	17.5

Source: Excerpted from "Self-Reported Comorbidities, Perceived Needs, and Sources for the Usual Care of Older and Younger Homeless Adults," by B. Garibaldi, A. Conde-Martel, and T. P. O'Toole, 2005, *Journal of General Internal Medicine, 20*(8), p. 728.

Table 4.3   Selected Self-Reported Important Needs of Homeless Individuals by Age Group

Needs Area	50 Years Old or Older	Less Than 50 Years Old
Finding a Job	43.2%	69.2%
Improving Job Skills	35.1	65.9
Having Steady Income	79.4	55.4
Relapse Treatment	32.4	51.4
Alcohol Treatment	21.6	43.5
Drug Treatment	32.4	47.9

Source: Excerpted from "Self-Reported Comorbidities, Perceived Needs, and Sources for the Usual Care of Older and Younger Homeless Adults," by B. Garibaldi, A. Conde-Martel, and T. P. O'Toole, 2005, *Journal of General Internal Medicine, 20*(8), p. 728.

Note: This table was constructed to depict the differences between the two strata in the population under study. The highest need from the self-reported data was by far housing, but the two groups were about equal in regard to it.

this knowledge, it may be that less-than-successful solutions would be proposed, and the information forces us to think about the subtleties of the situation.

Table 4.4   Selected Self-Reported Site of Medical Care for Homeless Individuals

Site	50 Years Old or Older	Less Than 50 Years Old
Community Clinic	64.2%	61%
Emergency Room	7.5	14.4
Hospital Clinic	7.5	10.9
Shelter-Based Clinic/Street Outreach Team	20.9	10.6

Source: Excerpted from "Self-Reported Comorbidities, Perceived Needs, and Sources for the Usual Care of Older and Younger Homeless Adults," by B. Garibaldi, A. Conde-Martel, and T. P. O'Toole, 2005, *Journal of General Internal Medicine, 20*(8), p. 729.

Note: In the original article, statistical comparisons were made for most of the variables studied.

6b. Collect and Array Data About Treatment Strategies

Collecting and arraying information about treatment strategies may seem to emphasize the roots of epidemiology for disease more than other fields in which we conduct needs assessments. How is this principle applicable to work in education, social service, government, business, and so on?

In an example in Book 1 of the KIT, a program was described that assisted borderline high school students in a suburban district to improve their grades in key subjects. The need was sensed rather than directly measured. Those in difficulty were recommended to see qualified substitute teachers who provided supplemental tutoring, instruction, and guidance. The program worked but not for the intended group. Failure in this needs assessment was apparent in not understanding who was not participating in the service. Analogous to the homeless situation here are the main findings from an evaluation of the educational program:

- The program was well intentioned but based on assumptions that should have been explored in greater depth (there is no substitute for probing into needs).

- Needs differ for specific groups and should be looked at that way rather than globally.

- B, B–, and C+ students tended to use the new service instead of those for whom it was intended—they wanted to get their grades up for college applications.

- Students with the most need did not utilize the assistance (they may not have seen it as helpful, or there may have been other issues).

6c. Deriving Recommendations for Action

Now the pieces are in place to make recommendations for dealing with gaps or needs. From Steps 6a and 6b, the NAC sees how many individuals require help, what services are there, and how these services are being utilized. In health, a lot is known about treatment strategies. If we find the cause of an outbreak of food poisoning, correcting the problem is obvious. While we cannot cure HIV/AIDS, we understand how to prevent its spread via safe-sex precautions. This is not to minimize difficulties in curing or stopping the spread of diseases.

In some areas of health there may be more determinism or known ways to attack causes of problems (e.g., a particular course of action will lessen the effects or decrease the incidence of a disease). In education, social concerns, poverty, and even business and industry, solution strategies may be more tenuous. How do or should we go about reducing pernicious problems—drug addiction, dealing with poor achievement in schools, resolving the needs of an aging population, beginning to turn around the degradation seen in downtown areas of cities, improving or maintaining market share, water pollution, and so on? Using epidemiology is recommended, but solutions may be less clear (see Example 4.2).

Example 4.2

Epidemiology Applied to Another Field

Everyone agrees that schools must raise the achievement of students. Various studies indicate that the United States does not fare well in science and math comparisons with other countries (Schmidt & Wang, 2002). State assessment programs and No Child Left Behind have put a harsh lens on student success. Resources are or should be available, but the amount of funding may not be commensurate (or may be grossly inadequate) for the need arising from raising the bar.

In 2006–2007, one state passed a bill that expanded the core of courses required in high school with more units of credit in mathematics and science. How could epidemiology fit in here? Consider the circumstances of a large urban district in the state. It has budgetary and staffing problems, and despite some improvement it is not yet up to par for state standards. To estimate

(Continued)

(Continued)

what the new law might entail, administrators decided to determine current student knowledge in various subjects. They had a wealth of data from a large database at the state level. The results from yearly state tests are published and can easily be linked to other variables and to data regularly collected and based in districts. Table 4.5 is an overview of the district's situation. It is a shortened version of district scores from the state.

Table 4.5   Current Percentages of Student Achievement in Mathematics by Grade Level (See Notes 1–3 Below Table)

Grade Level/Subject	District Results	Similar Districts	State Results
3rd-Grade Mathematics	53.2%	53.4%	74.1%
4th-Grade Mathematics	52.9	53.7	76.9
5th-Grade Mathematics	38.5	35.8	62.7
6th-Grade Mathematics	40.4	41.4	68.4
7th-Grade Mathematics	40.7	39.0	63.2
8th-Grade Mathematics	43.2	41.8	68.6
10th-Grade Graduation Test Mathematics Science	72.3 48.6	66.0 47.3	82.7 73.1
11th-Grade Graduation Test Mathematics Science	78.2 62.5	78.3 62.8	88.9 82.8

Source: Excerpted from *2005–2006 School Year Report Card,* Ohio Department of Education, 2006.

1. Test results are also available for other subjects (reading, writing, social sciences), but not all subjects are tested at all grade levels.

2. The standards to be achieved are 75% passing up through the 10th-grade test and then 85% for a district.

3. The database contains additional data for each district in the state. Examples are attendance and graduation rates, performance in certain subjects broken down by ethnicity, indices depicting the scores by performance level so that students below basic or at basic levels can be ascertained, percentages of core academic subjects that are not taught by highly qualified teachers, and trends in the district's achievement over a 4-year period.

While progress is evident, the district is far from achieving standards let alone offering more science and mathematics courses to meet the new core. To do so would necessitate major increases in funding. From this starting point, it might do the following:

- Build a table showing the assignments of certified math and science teachers

- Estimate the number of additional courses needed to meet the new requirement and how many certified teachers would be required and at what grade levels

- Compare the assignment and additional course tables to see how many teachers would have to be hired without adjusting for other factors (see below)

- Look at funding for the new core to see how many teachers could be hired

- Assuming that there is not enough money, consider adjustments to teachers' assignments and schedules to cover the gaps

- Think about hiring part-time substitutes or rearranging the schedules of lesser qualified teachers to respond to the problem (not the best of options)

- From the above work, formulate recommendations to meet the mandate (even to not be in compliance to see if the state would take any actions)

This example gives a sense of the difference between epidemiology in health and its use in another field. Look at Table 4.5. There is a remedial need indicated by the data; not enough students are reaching the standard. The state mandate for an enhanced core may not matter if large numbers of students are unprepared for the opportunity. For the district whose statistics are shown in the table, it was noted that 32% of the students fail to be promoted to the 10th grade (Sebastian, 2007). Hiring new teachers and offering additional courses in math and science are not guarantees of increasing academic achievement. There is a connection, but how strong it is and how much of the achievement discrepancy it will erase is more elusive.

Another, almost hidden need is imbedded in the problem. It raises questions about potential needs and solutions that can only be partially answered via a database. The district could examine state data about students at basic and below basic levels. This would give estimates of students not having the necessary skills, useful knowledge up to a point. How much remediation do they receive from tutors, in-class assistants, peer tutors, science and math teachers, and others, and how effective is it? With high numbers of students, what costs are necessary for remediation to enable them to participate in the expanded core? If teachers devote much time to remediation, what impact might this have on regular classroom instruction? How might it affect teaching those at more advanced levels in the subject matter areas of concern? Time is a finite resource to be allocated thoughtfully.

Coming back to the investigative purpose of epidemiology, data from teachers could be collected by surveys and interviews—self-reports about time and perceptions of the degree to which help for students was successful. These results would be incorporated into the supply and demand for teachers for the new core, the amount of remediation required, and solution possibilities. The problem is not simple, and the difference from disease to education (or other fields) is underscored. In some situations (lack of vaccinations for childhood diseases), the solution is obvious even when the cause requires additional investigation.

Highlights of the Chapter

1. A brief introduction to epidemiology and its purposes was provided.

2. The use of databases was emphasized and illustrated for childhood diabetes.

3. How databases are constructed, how data get into them, and problems in use were discussed. There are many ways to misinterpret results from databases.

4. The main steps for conducting an epidemiological study were briefly covered.

5. An adaptation of the generic process to an educational example was provided, as well as how epidemiology applications in other fields differ from those in health.

Special Note: Numerous good texts on epidemiology are available. Two, Bhopal (2002) and Hebel and McCarter (2006), will be mentioned, but many others are available. The two are excellent overviews of the methodology with detailed step-by-step procedures for epidemiological studies and are filled with well-explained examples.

5

Qualitative Methods for Phase II

Focus Group and Individual Interviews

❖ INTRODUCTION

Qualitative techniques such as community group forums, focus group interviews (FGIs), individual interviews, nominal groups, participant observations, unobtrusive observations, and others are routinely used in needs assessment. The focus here is limited to FGIs (two types) and individual interviews. What's behind that choice?

From the literature it is clear that focus group or individual interviews are preferred in needs assessments. When joined with surveys or epidemiological studies, they produce in-depth information about needs. The FGI can be done either as a cyber or virtual interview or on a face-to-face basis. The two versions may or may not produce similar results depending on the topic being investigated and the sampled groups. The author has been using these kinds of interviews as major ingredients in his research, evaluation, and assessment work. He is of the mindset that combining them with a quantitative method is a good route for needs assessment and yields a rich picture of a need area.

For example, Altschuld and Austin (2006) conducted an intensive FGI of a small national agency in which key staff identified strengths, problems faced by the organization, and areas for future attention. The FGI was part of a study in which Level 1 recipients of agency services were electronically surveyed. The two techniques together were much stronger than either by itself. The interview was an opportunity for staff discourse that did not take place in daily activities and was a cathartic experience they enjoyed and found of value. In advance of the face-to-face meeting, questions and topics were sent out to "prime the pump" for the interview. The impression of the needs assessors was that only several of the group members really had prepared for the FGI. Perhaps a little more prompting and reminders were needed.

Similarly, Kumar and Altschuld (1999) evaluated a complex teacher education program 2 years after completion for long-term impact with one method being in-depth individual interviews of students, faculty members, staff, and administrators. That information was supplemented by past reports, onsite interviews of graduates teaching in the general vicinity of the university, and their building administrators. Data were summarized, and the conclusions reached were quite different from prior quantitative looks at the program. This surprised project staff, and they took copious notes at a debriefing session of the qualitative study. The qualitative methods led to subtle, under-the-surface positive effects and ways in which the program was less effective than originally thought. Students were using program materials in a manner completely unknown to project staff, a finding that only the qualitative data unearthed.

In another project, a case study greatly enhanced a retrospective needs assessment of a national dissemination service. Evaluators/needs assessors went to four selected school sites to interview central office staff and teachers about access to science and mathematics teaching materials and the nature of educational reform in these areas and to look at records and reports available onsite (Cullen et al., 1997). This endeavor was done in conjunction with surveys distributed to national samples of teachers and administrators (see Chapter 2).

The results were striking and aligned with principles stressed in Chapter 2. First, levels in a system have different perspectives. In the interviews, administrators were familiar with national initiatives and had favorable views, whereas teachers were much less knowledgeable. For them, questions about the initiatives appeared to be in a foreign language. This was so pronounced that interview protocols had to be totally reworded, demonstrating the need for sampling multiple levels of respondents and the *within-method* variation (Chapter 3).

Second, the study underscored the use of multiple methods. National surveys indicated large discrepancies between the importance and the availability of technology and support in schools when the responses of teachers and administrators were compared. Level 2 personnel (teachers) saw much bigger discrepancies than did higher-ups. The qualitative work helped in explaining why, and "without going to ground" it would have been difficult to interpret and ascribe meaning to survey results. The interviews were essential for thinking through what was happening and for gaining deeper insights into the current and emergent needs of teachers.

Individual and focus group interviews were used in two of the studies in Chapter 2. In one they were a precursor to the development of the survey, and in the other they were integrated into the assessment effort. What are the steps that should be followed in a face-to-face FGI?

❖ STEPS FOR FACE-TO-FACE FOCUS GROUP INTERVIEWS

Step 1: Decide to Use the Technique

FGIs require a lot of arrangements to be successful:

- Is the FGI useful for our needs assessment, or would another qualitative method be better?

- Will FGIs produce the information needed, and are they worth the cost?

- Who should be involved, and are these individuals or groups accessible for the interviews (this may be easier said than done)?

- How will we arrange for the interviews and get people to them at the appropriate time and place (a problem for cyber focus groups as well)?

- How many interviews should be done?

- What questions should be asked and in what order?

- Will the questions require thoughtful rather than simple, not-in-depth answers?

- Who will conduct the interviews, and who will analyze the data?

- How will the data from the FGIs be tied into other needs assessment information?

Briefly examine the literature. Have FGIs been conducted for this area, and if they have, what did they ask and find? What can we learn that fits our circumstances?

Step 2: Select Individuals for the FGI

If the three levels of needs assessment have different perspectives, then do multiple focus groups but do not mix subordinates and super-ordinates in them. Some organizations are open and receptive to opinions and views, and a mixed group probably will work. Phase I should provide guidance for having this type of interview. But if the situation is closed and controlling, then subordinates may not comment or reveal feelings in a mixed group; an implied threat may be felt that curtails the honesty and openness of responses. If this is the case, homogeneous samples are recommended.

The other aspect of involvement pertains to Levels 1, 2, and 3. Although Level 2 is the most frequent target audience perhaps due to its availability and accessibility, consider conducting FGIs at Levels 1 and 3. There are several other features about sampling for FGIs:

- Replication is important.

- With replication and going to multiple levels, costs will rise.

- How many groups are feasible, and at what levels?

- The application of FGIs in needs assessment is very different from their application in marketing (see below).

FGIs may be used to understand how to sell products and services with participants receiving financial payment for taking part. The interview may be taped (video or audio for future use and analysis), and if a room with a two-way mirror is available (in specialized marketing firms), then the sponsor may watch people's reactions (body language, enthusiasm, facial expressions, etc.). Taping is only done with the permission of the participants.

For needs assessment, the circumstances are not quite the same. Issues might arise about the management of organizations, internal relations, or unanticipated problems that could be threatening. Therefore, follow standard procedures for the protection of human subjects and maintain anonymity in summarizing results. Videotaping might work, but if it would raise hassles, don't push it. Obtaining honest responses is always the main focus (see Example 5.1).

Example 5.1

Those Honest Responses

Yoon, Altschuld, and Hughes (1995; in Altschuld & Witkin, 2000) conducted several needs-related FGIs of international Asian students at a large university. The procedure was repeated with three groups, and in the third interview only female students showed up for the session. During the interview there was a radical, spontaneous shift away from the original questions that dealt with housing, academics, language, cultural adjustment, financial problems, and the like. The shift (probably due to the moderator of the interview being a former student and an international Asian female) was to sexual harassment ("being hit upon") by males within the community and external to it. Would such private and sensitive matters have come up if the interviews had been taped or if the interviewees had not felt that their responses were confidential and their privacy was fully respected and maintained?

Step 3: Attend to the Details and Arrangements That Make for a Successful FGI

There are critical things that must be attended to in order to have a successful interview. They require a lot more time than meets the eye:

- establishing criteria for the participant selection (direct service receivers, service providers, key administrators),

- mechanisms for contacting potential participants to see whether they are willing to be in a group,

- room location and time for the meeting,

- refreshments,

- comfortable seating in the room,

- a way to record the proceedings (a flash drive tape recorder would do), and

- having a cofacilitator and/or another person to take notes (so that the main facilitator will be able to concentrate on leading the discussion).

Obtaining an adequate sample for the interview may be a problem. In a study conducted by the author and colleagues (Altschuld, Ramanathan,

Ou-Yang, Barnard, & Holzapfel, 1996), college students who had graduated from a major urban school system and who were still in the area were needed for an FGI. An honorarium, travel expenses, and refreshments for a 2-hour Saturday morning session were offered. After extensive contacts, 10 students agreed to be at the interview, but only 2 came. The investigators subsequently learned that there were several meetings held later in the season where many of the students were naturally clustered together. It was through them that they mustered a full FGI complement (see Example 5.2 for doing something similar).

Example 5.2

Using a Conference for FGI Purposes

A center for public health preparedness wanted to evaluate a set of grand rounds lectures and awareness sessions it sponsored during the year. What was their quality and utility, and what topics might be offered in the next year?

A face-to-face FGI (one of three with the other two being via cyberspace) was used to ascertain these perceptions. This required the participation of public health practitioners who had attended numerous grand rounds events and were familiar with them. They were widely scattered, and having a face-to-face interview in a central location was problematic. Fortunately, a statewide conference was coming up, and many of the potential participants would be there. The enrollees for the conference were compared to those who had been to at least two grand rounds. A sample was selected, contacted, and involved in an informative interview during the conference.

The participants enjoyed offering their comments and views. The FGI was a change of pace, and they stayed for 15 min after the interview to continue discussing ideas and thoughts. Take advantage of situations like this. It is different from what participants would ordinarily do at a conference and perhaps refreshing for them.

Step 4: Determine FGI Questions

What questions are going to be asked, and how do they fit the needs assessment? In Example 5.2, the interview began by having participants briefly introduce themselves and their work and that of their agency. This activity gets people talking in the group and on a first-name basis.

Recall that earlier the use of the technique for a small national service provider was described. All individuals sampled for the interview knew each other, so another "icebreaker" was used. The question went something like this: "Think back to when you were younger (perhaps in high school or college) and tell us about something that happened to you—an embarrassing or crazy moment that occurred that you would like to forget. Let's have some fun as we get started." It and other questions were mailed a week before the session. The group definitely had a few laughs as the interview got underway. That first question helped participants relax in the group setting.

From there, move to specific probes about the need area under scrutiny. They are about things that cannot be easily garnered from surveys and databases and should require respondents to think and to provide explanations and descriptions, not yes or no responses.

- What are examples of this problem that come to mind?

- From your perspective, can you describe the severity of the problem?

- What factors do you think might be leading to or contributing to the problem?

- Describe the receptivity of the organization to resolving the problem.

- When you think about needs of the organization or actions that will improve the organization, what comes to mind first, and why?

- Describe how motivated individuals might be to work on or to resolve the problem.

- What might be indicators of this problem, and how can they be measured?

Have a few examples to prompt responses if the group is not forthcoming or if there is too long of a "pregnant" pause. The moderator should be alert and prepared to probe. In focus groups, unanticipated comments will be given that may necessitate further commentary. Be ready with questions like "Would you briefly explain that a little further?" or "I suspect that you had a little more to add in your response, so would you be willing to expand it somewhat?"

You want in-depth and considered responses with every interviewee contributing to the discussion. If there were six to eight key questions to ask of eight participants and you allotted several minutes per person for explanations and comments and several for a summary, the total time per question would be perhaps 20 min. Aside from beginning the meeting, six to eight questions may take up to 2 hours or more depending on what is being expressed.

That means that the moderator of the group has to keep it on task. In-depth responses are important, but if individuals run too long or are too cryptic, then the leader has to know when to push for more information or to gently curtail a person who goes on forever. Also, make sure your questions and the prompting of respondent thinking and answers will be important for understanding needs.

Exhibit 5.1 contains questions asked in the study of the national service agency (the interview was for 4 hours with a lunch break, more than twice as long as normal). Exhibit 5.2 is the cover letter sent to the group to set the stage for the interview, which dealt with evaluation and needs assessment. It was e-mailed 1 week prior to the interview. Exhibit 5.3 contains the questions for the evaluation with a needs orientation in public health (Altschuld, Yilmaz, Harpster, Pierson, & Austin, 2006; Yilmaz & Altschuld, 2008).

Exhibit 5.1   Sample FGI Interview Questions as Applicable to the Specific Concerns of the National Service Agency Needs Assessment

1. Warm-up question to get everyone relaxed and in the frame of talking.

2. For building organizational capacity, your funding has four main emphases: database management, program inventory/gap analysis, strategic planning, and expansion of membership. In terms of strengthening the organization, which emphasis do you see as most important, and what is the reasoning for your choice?

3. In your judgment, which area was *most* successful, and upon what evidence or data do you base your judgment?

4. By the same token, which area was *least* successful, and upon what evidence or data do you base your judgment?

5. Thinking about program inventory/gap analysis and strategic planning, what are some indicators of success of these emphases? How would you measure success?

6. What kinds of patterns in programs have you identified? Do some programs seem to work better than others? Please describe what your perceptions are.

7. What do you see as the most glaring need for the organization to attend to and improve upon its service? Why is it a need?

8. Assume that the federal government found a kitty of unexpended monies and would provide $100,000 for investment in only one of the four capacity-building emphases. Which one would you choose, and why is it more crucial or more needed?

9. If you were developing a needs assessment for the groups you serve to drive proposals for future grants and services, what would be some key areas to study?

   How would you measure these needs?

   What would be critical variables to be assessed?

10. Do you have any other thoughts or ideas to contribute that would be helpful or useful for the improvement of organizational capacity?

Exhibit 5.2   Sample Advance Organizer Letter Sent to Agency (Abridged)

Hi Everybody,

We are excited about joining you in February for what we are calling a Group Interview Plus. It is a combination of a focus group interview and a group discussion with some information for you to ponder in advance.

It is a face-to-face opportunity for us to get to know you and understand your perceptions of what has transpired as a result of your federal contract. Aside from that it is a unique time for you to share views with each other. We find that given busy day-to-day work, discussions like this don't often occur within organizations. So enjoy the interview.

There is some information given below to help you prepare for the interview, which will run from 10:15 to lunch and then reconvene for 60–90 minutes or longer if needed. The majority of the interview will be in relation to your federal grant.

How should you prepare for the interview?

- We hope that you will be very open and frank in your comments during the interview. All responses are important, and there is no right or wrong answer.

(Continued)

(Continued)

- The first question (and there is a reason for it) asks you to briefly describe something about yourself that no one else knows or a ridiculous situation in which you found yourself at one time or another.

- Since there is a fun dimension to this question, please keep what you are planning to say secret until the interview. Thanks for cooperating and for your sense of humor.

- Think about the array of services provided to members and what you perceive to be member needs. What do your members/partners require to achieve better outcomes?

- Next, read through the national proposal and think about its main emphases. What do you consider to be the highlights of what has been done so far, and what probably should be strengthened?

- Then back up and reflect upon the reasons behind your answers. Is there any evidence you can cite or that may be guiding your thoughts?

- Jot down some general notes about the above questions and bring them with you for referral as we progress through the session.

- Also consider other things that would be helpful to the organization as it seeks to improve its capacity and effectiveness.

We will see you in February and look forward to a positive and enlightening discussion.

Exhibit 5.3   Focus Group Interview Questions as Used in Public Health

1. Warm-up:
   For us to get a sense of one another, please provide your first name, what your position is, and something about yourself that is unique and no one knows.
   Thanks! That was a lot of fun, but now let's move on.

2. Take a moment to think back on the public health grand rounds (PHGR) events that you have previously seen and/or with which you have interacted. Briefly describe their impact on you (changed my thinking, began to look at or explore a new topic, etc.) and how you may have used them (encouraged others to view them, incorporated them into a class, etc.). Note: This might be separated into two questions.

As participants present their ideas and thoughts, look for the comments that might need clarification and ask for same.

Also look for trends in the comments, summarize them, share the summary with the group, and see if they agree with your characterization.

3. What you have described has been interesting and informative. Reflecting on the array of ideas just given, are there other ways that the PHGR events might be utilized?

Summarize in a manner similar to the previous question.

4. While we know that the PHGR events have been incorporated into group sessions and in classes, the majority seem to have been viewed by an individual attending the session or watching the event later alone. In what ways could we get more viewing and use by groups?

See above.

5. Currently we advertise the PHGR events by electronic announcements and in our electronic newsletter. These work well, but we are interested in obtaining a wider viewing audience, especially one that would be doing so in some kind of group endeavor. What suggestions would you have for us to do so?

See above.

6. Choosing topics for PHGR events is always a sensitive type of undertaking, particularly when the goal is to attract widespread participation in the events. Please share with us your thoughts on topics that would tend to attract a larger audience.

See above.

Express thanks for their involvement and ask them to complete the general information questions.

Across the exhibits, there are common themes pertinent to needs assessment:

- Each interview starts with a way to get people at ease in the setting.

- FGI questions contain general guidelines or prompts for the facilitator to use if necessary. They are only for the facilitator.

- Questions that focus on needs may refer to them in an indirect manner and may reveal more about needs than direct questions.

- The interviews do not consist of many questions as time is limited.

- Notice how it is possible to get into detail in the interview. The concept is to "drill" down into the topic.

These guidelines are offered with the idea that the questions are tailored to the issue of concern and the context of the needs assessment. Although not described above, the interviews were led by a facilitator with an assistant facilitator or another person taping the session and taking notes.

Step 5: Conduct the Interview

Since many of the aspects of FGIs have been discussed, the stress is on a few keys that make for success:

- The facilitator should have familiarity with the questions and their sequence before the interview rather than reading each one before asking it (this can turn off a group).

- Don't cue participants verbally or by nodding one's head for there are no correct answers to questions.

- Be prepared for answers "off the cuff" that might be important and be ready to follow up on them.

- If participants springboard (branch off of the response or comment of another participant), allow this to happen and see where it goes.

- Gently prod those who don't provide much depth to say more and those who go on too long to summarize.

- Frequently recap as a way to keep the interview on track, make sure that it includes majority and minority views if applicable, and ask the group if it is a fair assessment of what has been said.

- The use of a second or cofacilitator is encouraged.

Step 6: Analyze and Report Results

Books 1 and 4 in the KIT contain procedures for looking at qualitative data, so here just a few points are highlighted. You have to collate and pull together the thoughts and ideas from open-ended comments into coherent themes that provide valuable information for the needs assessment. At the same time, you have to be cognizant of what other methods have produced. The qualitative techniques do not lend themselves to discrepancy analysis. They have to be blended with other findings into a comprehensive and intelligible picture about the need area(s).

Begin by debriefing with the cofacilitator. What impressions are there, what stood out, and what seemed to be emphasized? Then, each person individually summarizes from reviewing the tapes (transcripts) and notes that were taken on a question-by-question basis. Consider what terms or main ideas were repeatedly stated or mentioned (these are variables), the threads or themes to the discussion of a question, places where the participants disagreed or fell into different camps, and themes (overarching ones) that seem to span answers to individual questions.

The analysis process is funnel-like in nature. It starts with the variables that underlie the data and moves to a smaller set of overarching explanatory themes as depicted in Figure 5.1. Once this sorting process is done, the needs assessors see how the results relate to those from other methods.

After analysis is complete, a full report of interviews should be generated and available for those interested in seeing how the procedure was conducted. Most important, the focus has to be on usable findings for enhancing understanding of needs and possible actions that might be taken to resolve them. A practical procedure might be to provide tables that communicate the variables uncovered and the themes—especially overarching, explanatory themes. Tables 5.1 and 5.2 are examples of these. They were developed for the public health project.

Figure 5.1   Schematic for Dealing With Qualitative Data in Needs Assessment

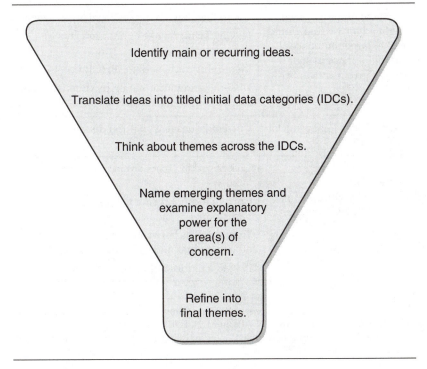

Identify main or recurring ideas.

Translate ideas into titled initial data categories (IDCs).

Think about themes across the IDCs.

Name emerging themes and examine explanatory power for the area(s) of concern.

Refine into final themes.

Table 5.1   Themes by Question

Question	Themes
Why have you attended PHGR events?	Interesting seminar topics Supporting PHGR activities Updating skills and knowledge about the public health issues Meeting public health professionals
What kind of incentives might be considered for attending PHGR events?	Free food Offering academic credit
What are your impressions of the quality of PHGR events?	Well-prepared speakers Good handouts and presentations
What have been the effects of PHGR events on you?	Enhanced knowledge about the PHGR topics Increased awareness about the professions and people that play roles in public health Practical application ideas
How could we get more viewing and use by groups?	Disseminating the information by various channels such as Web sites
Are there other ways we might advertise that would lead to larger audiences, especially those in some kind of group endeavor?	Contacting different professional organizations Having faculty make announcements in their classes Identifying audiences for PHGR events Disseminating information by different tools (newsletters, e-mails, Web sites)
What topics might attract a larger audience?	Current issues (e.g., natural disasters)
Are there any time or location issues that prevented you from attending PHGR events?	Lack of parking space around the current location Current location being too far for some participants Time conflicts between PHGR events and other activities and class sessions PHGR activities not being in the morning

Source: From *Summary and Analysis of the Focus Group Interview Process: Evaluation of Public Health Grand Rounds,* by J. W. Altschuld, H. B. Yilmaz, A. Harpster, B. Pierson, and J. T. Austin, 2006, a report for the Ohio Center for Public Health Preparedness, The Ohio State University–Columbus.

Table 5.2 Themes Across Questions

Question Area	Themes
Why have you attended PHGR events? What topics might attract a larger audience?	Current issues such as pandemic flu, climate change, and agro-terrorism are important topics
What are your impressions of the quality of PHGR events? Are there any time and location issues that prevented you from attending PHGR events?	Current location is not appropriate for PHGR events
What kind of incentives might be considered for attending PHGR events? Are there any time and location issues that prevented you from attending PHGR events?	PHGR schedule could be changed
How could we get more viewing by groups? Are there other ways we might advertise that would lead to larger audiences, especially those in some kind of group endeavor?	Getting more faculty involvement is necessary Information should be disseminated by various channels Related organizations should take active participation in the process

Source: From *Summary and Analysis of the Focus Group Interview Process: Evaluation of Public Health Grand Rounds*, by J. W. Altschuld, H. B. Yilmaz, A. Harpster, B. Pierson, and J. T. Austin, 2006, a report for the Ohio Center for Public Health Preparedness, The Ohio State University–Columbus.

❖ STEPS FOR CYBER OR VIRTUAL
 SPACE FOCUS GROUP INTERVIEWS

In the face-to-face FGI, getting people to one place at a precise time for the session may be difficult. They have busy lives and schedules and may have to make numerous arrangements to come to the session. If something goes wrong or unexpected events occur, they may not show up even when they have made a commitment or there is a monetary incentive to do so. Sample loss will occur. With small numbers (7–10), a few individuals not showing up affects the interview and its results. This is not so much of a concern when group members are at a conference or when they are housed in physical proximity as in a manufacturing plant or in an agency. The FGI can be arranged in the course of normal events.

Now consider an agency, a bureau of state government, or a business with geographically dispersed branch offices. Convening a face-to-face group would be inconvenient and expensive. An option would be to conduct it via the Internet. Everyone doesn't have to be in one location, but all participants do have to be online at the appointed time. This could be done in several ways. A *synchronous* FGI is where participants are online at one time with a facilitator leading the group. An *asynchronous* version is like a "chat room." Respondents supply responses to questions after which they are able to see what others have said, thereby allowing them to springboard or add other thoughts to their comments. Since asynchronous interviews do not take place at one time, the FGI would not be led by a facilitator and could be conducted over a week or similar period of time. A *combination of approaches* is possible, with some participants being interviewed in real time and others when schedules permit, but their comments are subsequent to those in interactive sessions.

The synchronous FGI requires a great deal of planning and thought. Potential group members have to be contacted, and a convenient time for the Web-based meeting must be established. For the asynchronous FGI, only commitment to participation in that week is required. Of course, the latter loses some of the spontaneity of its interactive component. The choice depends on the sample, timing, and commitment.

In cyberspace, the role of the leader of the interview is not as clear as it could be. For example, one cannot see the facial expressions and body positions (unless by videoconferencing) of respondents. You will be less able to "read" the group, to know when to probe and when to push in regard to springboarding. This interview may not generate as much information, but when everyone cannot be in one physical place, this is the only viable option. The loss of some probing might be tolerable.

The Web-based FGI has a nice feature depending on the software employed. Each person in the interview may see the comments and thoughts of others on his or her monitor, and the leader and coleader can do the same on a big screen or a wall. They can immediately review the responses and may detect trends more easily. Beyond that there is no need for transcription since comments are automatically recorded in a running record by question and respondent. This makes for faster processing of the data.

But a subtle problem may occur when respondents can see each other's responses. In one case, the author noticed a form of "group-think" with respondents beginning to cite or even adopt the words and statements of other group members. In a face-to-face interview, respondents hear each other's comments, but they aren't prominently displayed in front of them. The latter may lead to more echoing of what others have said. The cyberspace technique might work better if there was a delay before the responses showed up on an individual's monitor. This would tend to reduce parroting. Also, typing and thinking speeds are variable, causing lag time for some and boredom for others. In cyberspace, everyone is answering at the same moment rather than one-by-one in a group. A slow typist may cut short responses just to keep up with others. This would not be part of a traditional FGI.

The literature about cyber FGIs is emergent. (See Adler & Zarchin, 2002; Burton & Bruening, 2003; Greenbaum, 2000; Kenny, 2005; Schneider, Kerwin, Frechtling, & Vivari, 2002; Turney & Pocknee, 2005; and Yilmaz & Altschuld, 2008.) It is not copious with still much to be done. Topics such as how groups should be engaged and facilitated, the role of advance organizers, or how to get expansive thought and commentary going are not explained or described in depth. Nor is the concern with groupthink or response times just described. Most of the steps required for the traditional FGI equally apply here.

Step 1: Decide to Use the Technique

This may be the only method if the participants are geographically dispersed!

Step 2: Select Individuals for the FGI
(See This Step for the Traditional FGI)

Step 3: Attend to the Details and
Arrangements That Make for a Successful FGI

Details regarding locating and using the appropriate software need attention. Many universities, state agencies, businesses, and organizations have the capability for such interviews. Scout out what is available; look carefully at technical issues and what might be needed for a good session. What costs might be entailed for the interview? Will technical help be available? Compare costs with the budget and determine the best way to proceed.

**Step 4: Determine Interview Questions
(See This Step for the Traditional FGI)**

**Step 5: Conduct the Interview
(See Previous Content in the Chapter)**

Step 6: Analyze and Report Results

This process is like procedures for a traditional FGI, but note responses are accessible as soon as the interview is over. The author and colleagues (Altschuld, Yilmaz, et al., 2006) led two cyber interviews and a traditional one on the same topic, thus providing a way to compare the two versions of the technique. Results were similar with somewhat more in-depth comments coming from the face-to-face meeting (Tables 5.1 and 5.2). There seemed to be less concentration on the part of participants as the virtual group interview was coming to an end. Repeating and contrasting different forms of the interview are helpful for thinking about the quality of the findings.

❖ STEPS FOR IMPLEMENTING INDIVIDUAL INTERVIEWS

Individual interviews are a good choice for the qualitative part of a needs assessment, but generally they don't directly produce discrepancy data; rather discrepancies are inferred. While this is a truism, there are exceptions. In public health (PH), the concept of after-action reporting is routinely employed to understand problems encountered with assistance. After an event (disaster, epidemic) has happened and is abating, deliverers of services are questioned about preparedness for serving those in need and/or how well the services worked. By this means, it is possible to get a feel for discrepancies that, if tied to other data, can be informative of needs—especially training needs.

Hites (2006) looked at the training of PH workers who went to New Orleans in the aftermath of Hurricane Katrina. Due to minimal resources and time, he had 3 days to interview PH workers as they were actually providing help and support to people in the Gulf of Mexico area. Hites asked questions such as the following:

- What have you been doing here to help victims of the hurricane?

- What is your background and training (what kinds of specialized training do you have)?

- Have you had to do things for which you were not prepared? If so, what, and how did you handle the situation?

- If there were another disaster like this, what would you recommend for training of PH workers to prepare them for what they might encounter?

- From a personal perspective, what kinds of problems have you run into (fatigue, mental health, personal hygiene, morale, etc.)?

- Describe any ways in which we could improve similar situations in the future.

While retrospective recall may affect responses, they are still helpful in determining what might be missing in prior training or, better yet, what might be included in subsequent training for help providers. Data from records about training (when it was received; its intensity, quality, and appropriateness) round out what is being gained from the interviews. But even with that information there is no substitute for the firsthand perceptions of those on the front line. The emotional intensity in the interviews adds another piece to the puzzle of understanding. Quantitative data simply cannot capture this human perspective.

(Parenthetically, a close friend works for the Red Cross and was on location delivering services for 2008 Hurricanes Gustav and Ike. Many of the issues explored by Hites [2006] are similar to the content of the friend's e-mail messages. This is also reminiscent of interviews conducted at Ground Zero shortly after 9/11. A group of doctors and nurses went to the site after their hospital shifts to volunteer assistance. They were shell shocked and stunned by the enormity of the tragedy and how few survivors they were treating. As they were being questioned by a television reporter, their anguish and the toll it was taking on them psychologically and physically were obvious. The former seemed much more devastating than the latter.)

Pertinent to this discussion is the Targeted Capabilities List of the Department of Homeland Security. It includes estimates of how effective specialized emergency teams brought in to help with a disaster would be. Since they would be operating at a frenetic pace over an extended period of time, what toll would that exact, and what would happen to effectiveness? How long could they sustain their efforts? What kinds of support would be necessary to keep them at reasonable efficiency? What might be their emotional states, particularly if the devastation was great? Would the affected sites be prepared to take full advantage of the specialized expertise they would bring? What kinds of supplies would be required for maintaining levels of service? What might be the consequences if, say, two or more Level 5 hurricanes were to hit in a short time hundreds of miles apart as occurred with Gustav and Ike? Would we be ready for this?

Hites's (2006) work was followed by a more intense, similar project by the North Carolina Center for Public Health Preparedness led by Davis (2006). Her team, with more staff and resources, conducted an in-depth investigation of what was happening with the itinerant PH workforce that was coming onsite. The study produced a lengthy report regarding needs for preparedness.

Investigations like these could also be extended beyond workers (Level 2) into the responses exhibited by organizations (Level 3). How did they coordinate their response? How well prepared were they at an organizational level to offer assistance that was quite a distance away? What kinds of problems (logistics, drain on ability to serve their home areas, reductions in local supplies, etc.) did they run into, and how were they handled? What lessons did they learn from the experience? What would they do differently now that they have gone through the situation? What are their strengths and deficiencies? The importance of the information coming from interviews like these cannot be underestimated. Again, it is doubtful that what is really taking place is fully understood without data from them.

Interviews are typically used in three ways for the purposes of needs assessment. *First,* they provide guidance for the design and implementation of surveys. One of the studies in Chapter 2 used them this way. *Second,* they are incorporated into the assessment process as an integral part of the complex mosaic of needs assessment data. They are not an adjunct to another method but have major import on their own and are critical input for comprehending the nature of needs. *Third,* after quantitative data have been collected and analyzed, experts and/or others who have reviewed the data, participants answer questions about the interpretation of the findings.

All three of these are good strategies for interviews in needs assessment with one concern thrown into the mix. It is that the method is just too important to be viewed as adjunct to another—its information yield can be great. Below are specific steps regarding the conduct of interviews for maximum impact on the study of needs.

Step 1: Decide to Use the Technique

As with any method, consider its overall role in the needs assessment process because much time and resources could be required. If interviews are intensive, an interviewer will probably do a maximum of three per day. If more are done, fatigue enters in, and subtle cueing of interviewees may take place. If numerous interviewers are needed, they will have to be trained so there is consistency in the process. Interviews are usually taped (in one international study in which the author participated, this was not permitted) and then transcribed before going through an analysis procedure. If interviews are to be done for each of the three levels in needs assessment, the costs of data collection will rapidly escalate, so make decisions after careful deliberation. If only a few quick, not-as-scientific interviews are appropriate, then the decision will be relatively easy.

Step 2: Select Individuals to Be Interviewed

When all three levels are to be interviewed, it is likely that there will be different questions for each, a *within-method* variation. Also remember that interviews are a personal encounter between the interviewee and the interviewer. They are an opportunity for an individual to offer opinions and perceptions and expand on his or her thoughts. They enhance a sense of involvement in the needs assessment process for the three levels, especially Level 3. Level 3 may not be too prominent in surveys and/or FGIs, so this group may jump at the chance to be interviewed.

Step 3: Attend to the Little Details and Arrangements to Make the Interviews Happen

Once interview protocols are developed, the implementation details are less cumbersome than for FGIs. Interviewees are contacted, and arrangements are made for time and location. Since most interviews are taped, this should be mentioned during the initial contacts. To ensure that the interviews are as free from bias as possible, all comments are treated as confidential and in a manner that will not identify the person providing information. This is particularly important for individuals from Levels 1 and 2. Explain the purpose of the interview, emphasizing that all responses are confidential and anonymity will be maintained in results.

Step 4: Coming Up With Interview Questions

Questioning is questioning is questioning. There are similar principles to follow whether a group or an individual is interviewed with perhaps one major difference. In an individual interview, more questions can potentially be asked, and more probing into responses can take place. Start with an icebreaker after initial cordialities. For example, people like to talk about what they do, so begin by asking something like this:

"Now that you know the focus of this interview, could you briefly describe aspects of your job that are related to that focus?"

That query could be followed by things like this:

"In regard to your response, what worked well, and how do you know it worked well?"

"As you know, it is not a perfect world. Are there things that could have been done better or problems that you encountered? Please explain them."

"Aside from what you personally do, what do you think are the organization's strengths for dealing with the area of focus (or issue), and what might it improve upon in this regard?"

There are many other questions earlier in this chapter that could be adopted for the individual interview. Consider their use with individuals.

In needs assessment, one may have to push respondents to talk about concerns that they may be reluctant to discuss with someone from outside the organization (if pertinent) or with whom they are not familiar. They may not be willing to open up about what is not going well. The organizational culture may not support an honest and frank dialogue. So why should they all of a sudden become free with what they are saying? From this perspective, note the phrasing above: "As you know, it is not a perfect world." The choice of words is deliberate. The idea is to frame the question in a way that makes it reasonable for the person to talk about and relate to the nature of concerns. Without this type of wording, socially acceptable answers might be observed more than those that get more into the nature of needs.

Altschuld and Thomas (2009, in press) found that wording survey questions to prompt more realistic and honest responses changed the results of a national survey. Respondents were cued that providing descriptions as to what was really happening (not what the organization wanted to hear or playing the game) was acceptable. The instructions dramatically affected the outcomes, and the same reasoning applies to interview questions.

Step 5: Conduct the Interview (See Previous Content)

Step 6: Analyze and Report Results

Individual interviews will adhere to most of what has been said before about analysis and reporting of results, so look for trends in the interviews and overarching themes! Occasionally comments and serious issues are brought up off-the-record. (This occurred several times in the author's experience.) They are not part of the summary of the interviews, and the privacy of the respondent must be respected.

Highlights of the Chapter

1. A rationale predicated on observed needs assessment practice was provided for the focus on only three qualitative techniques as opposed to a broader array of methods.

2. The techniques were presented with an emphasis on how they might be used in needs assessment. Great attention was paid to the traditional FGI and then generalized to the cyberspace and individual interviews.

Note: Throughout the chapter, examples were imbedded from real-world needs assessments.

6

The Other Parts of Phase II

Causal Analysis and Prioritization

❖ AT THIS POINT IN PHASE II

The needs assessment committee (NAC) has a lot of information about needs for which the organization might invest resources and time to resolve underlying problems. Now it is necessary to determine what is causing the discrepancies and to prioritize them. Which of these activities should be done first?

From one viewpoint, it would be better to know what causes the three or four seemingly most important needs before priorities are set. How correctable are the problems, can they be handled by the organization, would there be sufficient motivation and resources to work on the deficiencies, what kind of disruptions would occur by shifting direction to several key needs, is there a good chance of resolving them in a successful manner, and so forth? It would be good to have answers to these questions before prioritizing.

But why not just set priorities and then look at causal factors? If five or six needs were apparent and two were seen as more important, all that would have to be done is to look at what causes them instead of having to cope with the whole set. The choice is left to the NAC and the facilitator since they know what plays best in the local context. Because there is no real response to the rhetorical question, arbitrarily the discussion will start with causal analysis.

❖ CAUSAL ANALYSIS

Step 1: Capitalize on What Has Been Done Before

Go back to Phase I and initiating the process. Recall that the NAC in its deliberation began to build some very utilitarian tables and summary devices. One of these is shown below (Table 6.1). It illustrates what is available for the committee. Additionally, in needs assessment tables like this are usually updated as the committee goes about its work and have many details in them. Pick key tables and summaries and briefly reexamine them.

What have we learned about causes? What is understood about them? If the tables are dated, they allow the committee to track the narrowing (i.e., the reduction in the scope of focus).

Step 2: Pose Some Basic Questions

Ask some basic questions to ensure that unnecessary activities and expenses are undertaken or incurred, respectively, and to get the NAC on the same page.

Table 6.1   A Phase I Decision-Oriented Framework

Need Area and Subareas	Further Actions Required	Reasons for Further Action	Preliminary Causes	Ideas About Solutions
Area 1 Subarea 1 Subarea 2				
Area 2				
Area 3				
Area n				

From *The Needs Assessor's Handbook*, by J. W. Altschuld and D. D. Kumar, 2009, Thousand Oaks, CA: Sage. Used with permission.

- What do we know about the causes of the needs that we have been investigating?

- Has the needs assessment process moved to a small number of needs so that formal prioritization may not be required?

- What in terms of causal analysis and prioritization would help decision makers to comprehend why certain courses of action are being recommended to them?

- Would arguments to do so be compelling?

- Do we feel that we are in consensus about causes and priorities?

- Are we certain that the highlighted needs should be resolved and become the foci of organizational efforts?

- Is our recommendation based upon the needs assessment that needs do not exist or are of insufficient importance to proceed any further?

The discussion of these concerns should not take much time. The group has seen all of the relevant information before and should be primed to consider what it wants to do next. Assume that the committee wants to explore causality further before going to prioritization.

Step 3: The Details of the Analysis

The level of detail in causal analysis depends on the complexity of the need and the information available to the committee. One suggestion is to start with a relatively simple but effective procedure—the fishbone diagram.

This is one of the techniques mentioned in Book 1 of this KIT. The others were fault tree analysis (FTA) and cause consequence analysis (CCA). FTA is the most in-depth and probing of the causal methods, but it comes with the costs of time and being somewhat difficult to apply. CCA is a quick method for looking at cause, which as you would suspect suffers from a lack of detail. Another technique that might be considered is quality function deployment, which has elements of causal analysis imbedded in it (Altschuld & Witkin, 2000).

The fishbone diagram fits midway between FTA and CCA. Even though its main purpose is for quality control, it is in all probability more well known to a wide audience of practitioners who work on problems and needs in society and business contexts. FTA originated in engineering and is more established there than in the social realm, but there are some (not many) relevant applications in other areas (Altschuld, 2004).

3a. Creating the Fishbone Diagram

This can be done with a small or moderately large NAC. To generate a fishbone diagram, select a need or a discrepancy that then becomes the head of the fish as demonstrated in the generic diagram in Figure 6.1. An enlarged version of this is placed on a wall or a blackboard for the entire group to see. All subsequent work of the group is entered onto it.

The head is the need with a spine or backbone extending to the left. Off of it are major bones labeled Workers, Methods, and so forth. The technique comes from quality control—hence the terms assigned to the bones.

Figure 6.1   General Fishbone Structure

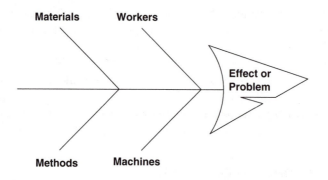

Source: From *Guide to Quality Control,* by K. Ishikawa, 1983, Tokyo: Asian Productivity Organization.

The next step is a brainstorming exercise in which NAC members suggest causes related to each of the major bones. There is a quiet period before this when they look at the diagram and briefly write down possible causal factors for each bone. After 10 min or so (when the group is not jotting down many ideas), the facilitator asks for those causes by going around the group in a round robin fashion. This is done major bone by major bone. The process continues until a fairly complete diagram is produced, which is evident when few new causes are being proposed (the group is exhausting its thoughts). In some if not most cases, it will be necessary to consolidate causes that are similar in nature. It doesn't take much time to do this.

Then the task before the NAC shifts to selecting the most likely causes of the problem in the fish head. This can be accomplished in several ways:

- Individuals could be asked to pick what they see as the top five causes of the problem.

- Each person could rank-order his or her top five choices.

- Some version of a zero sum game could be used (you have $10 to spend and should distribute that money in whole dollar amounts to the causes that you think are most prominent).

After this is done, the facilitator goes around the group and tabulates the responses of individuals directly on the diagram. This gives a picture of which causes were perceived as contributing most to the problem. Figure 6.2 is an example of the overall approach as used to ascertain what was causing the problem (or need) of bad-tasting coffee. The causes that seemed to be affecting this need for the group are circled on the diagram.

From there the NAC engages in a discussion about how well this process worked and whether it felt that the pinpointed causes were accurate. This discussion is important for understanding the perceptions of the group and the basis for its members' ideas. There is a fact here that is so obvious that it may escape notice and bears mentioning.

The NAC is not your ordinary group. Its members have been carefully chosen from the get-go as to what they can contribute to the overall assessment of needs. They are knowledgeable. They have studied needs for an extended period, and they already have a great deal of insight into the problem. Stated succinctly, they constitute an especially well-informed group. The fishbone diagram they produce will tend to be based on reasoned and documented information and should be highly utilitarian.

Now as they are in final discussion, note what they are saying and keep a record of it. It may also be wise to have them think about the fishbone over a week and respond then as to whether they still view it as useful for organizational decisions. The amount of time necessary for conducting a fishbone session is perhaps several hours, but the session may go faster if the number of group members is not too large. Once there is agreement on the diagram, it should be dated for inclusion in the needs assessment report.

The fishbone diagram may be put out for review, examination, and input from a wider group of organizational personnel, but exercise some caution when doing this. Needs assessment carries with it a political connotation and may involve a change in the direction of an organization and its use of resources. If the posting is in any way premature, it

Figure 6.2 Fishbone Diagram for the Problem, "Bad-Tasting Coffee"

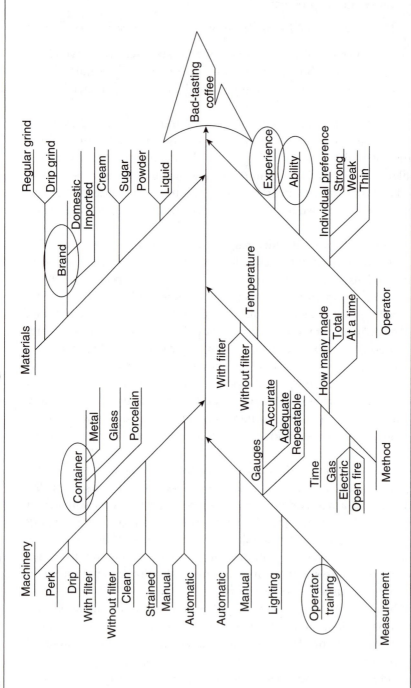

Source: From *Gone Fishing*, by M. Jones and K. Limes, 1988, unpublished manuscript, The Ohio State University, Educational Services and Research Department, Columbus.

Note: Circled entries are major potential causes highlighted by the group.

could lead to acrimony. It would be better to do this when the diagram is in relatively final form that is agreed upon by key parties and there is a strong rationale underlying it.

3b. A Few More Features of Fishbones—
What Can You Do With the Methodology?

The fishbone approach has another advantage over FTA and CCA. Aside from straddling the gap between them and the high yield of detail for the time expended, the diagram can accommodate looking at problems in processes or flows of work. In this instance the diagram is a little different (a more linear visual display) than what was described above. The head remains as previously, a problem or a gap. The change is in the body of the fish in that it is a sequence of steps or processes that could be causing the problem. A process could be failing or not working optimally.

The task for the group is to determine factors leading to deficiencies in each of the processes. The NAC goes through the same general fishbone strategy but looks at each process and what could be causing failure in it. This kind of fishbone is depicted in Figure 6.3. The NAC produces what are basically separate fishbone diagrams for each part of the process.

In sum, the technique seems very fitting for assessment of causality in needs assessment. But one concern is that the language and terminology come from the domain of quality control. It has a manufacturing ring to it, so you may want to use different terms. Even though it is an adaptable technique, there is a need to make the transition explicit. What might be done for socially oriented needs that are less mechanical in nature is to attach different labels to the major bones.

Consider a fishbone for children not performing well in school. Low achievement would be the head of the fish, and then coming off the spine could be categories such as the following:

- health and related matters;

- the nature of the home environment and how it does or doesn't support the efforts of the school;

- sports and other competing activities that may take away time from study and academics;

- resources available at home or in the school;

- community values as related to education and how, in turn, they permeate the milieu of the school; and

- the educational culture as embodied in expectations for achievement and administrative policies.

Figure 6.3 Fishbone Diagram for Determining Causes of a Production Defect—"Steel Pipe Scars"

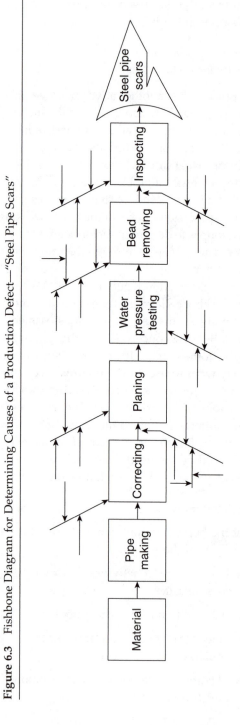

Source: From *Guide to Quality Control*, by K. Ishikawa, 1983, Tokyo: Asian Productivity Organization.

As you adapt and modify the technique to fit different types of problems, other things may arise. For example, the third point above might suggest that two fishbones be developed—one for males and one for females. Some of the causal factors might not be the same or may vary in intensity for the two groups. The flexibility of the technique is only limited by the imagination and inspiration of the NAC and its facilitator. Use it as best suits your situation.

The last thing to enter into the mix is that there is even more information readily accessible about causes and solutions that might deal with them. Some of the earlier materials generated by the NAC contain ideas about solution strategies. These should be brought to the fore and compared to the main causes now identified by the NAC. Briefly glance over them to see what their features are and how they align. Probe into them by asking:

- From an overall viewpoint, how well do they fit the causal elements?

- Do they deal with some causes better than others?

- Could they be consolidated with other solutions to deal with the majority of causes?

- How complex are they, and in turn how acceptable would they be to the organization?

- If complex, is it possible that the solution could be divided into parts that could be more easily implemented and that would be more acceptable to the organization?

This thinking would be useful in regard to prioritizing. Indeed some prioritizing procedures include aspects of solutions along with criteria associated with the importance of a need. A need that demands a complicated and involved solution may be prioritized lower than one that can be simply resolved; hence the logic for the last question is revealed.

Going further, if the assessment is being done across organizations, then the criteria of ease of solution will have high priority. A multi-organizational assessment tends to work best when it identifies needs that can be resolved in a relatively quick manner. Rapid success fosters more cooperation from collaborating organizations and hopefully will propel later work on more intricate and involved problems. Starting with complex needs could be discouraging and lead to less-than-enthusiastic support for participating in joint ventures—they are just too much at the beginning of the cooperative venture.

By doing causal analysis and by going back to earlier aspects of the needs assessment process, the NAC will have more understanding of how to prioritize. The time is well spent.

❖ MOVING INTO PRIORITIZATION

Step 1: First Considerations

Prioritization is already underway as implied. The NAC has consistently been reducing the focus of its effort. It must limit the number of needs that require organizational attention; there are almost always too few resources that if spread thin may not produce much meaning and change.

Organizational energy is in danger of being dissipated. We are unable to deal with everything; we have to narrow in and feel that something will be accomplished and achieved. The reduction in scope is neither negative nor nefarious but a fact of life in needs assessment. Eastmond, Witkin, and Burnham (1987) cautioned that if an assessment is too broad in focus, it will not produce much of value. (See Book 2 of the KIT for numerous ways to get the assessment started with a focus on a more limited set of concerns.)

Step 2: Look at the Kind of Needs to Be Prioritized

Because prioritization can be simple or complex, preliminary thinking is in order about the needs that have been highlighted and their causes. What are they like, how involved are their solutions (at least from what we currently know), and so forth are typical things in front of the NAC. If it is not going to be difficult to come up with priorities, then use straightforward procedures for establishing the top needs for organizational attention. If it is going to be difficult and particularly if the situation is contentious and priorities will be challenged, a more explicit and detailed procedure will be required.

Step 3: Simple Prioritization

3a. Rank Ordering

When the needs are not many (15 or fewer), have group members rank them. It is useful to provide some general criteria or guidance for ranking along the lines of the following directions:

"Put the needs into rank order based upon your perceptions of what are the most critical or important needs for the organization to address and invest its resources into alleviating."

or

"Put the needs into rank order based upon your perceptions of what are the most critical or important needs for the organization to address currently, to use its resources for alleviating, and for which it would have the greatest chance of success."

Both prompts have several criteria imbedded in them. Make these explicit so that the NAC will have some basis for discussion if there are major differences in the rank orderings of its members.

Record the rankings and obtain average rankings of the group. Report the average as well as the pattern of rankings. Put the needs into a table starting with the one receiving the highest average rank. For the most part, it is anticipated that there will be general agreement in the NAC. Priorities will be clearly delineated. If disparities arise and are evident, use them as a discussion point for the group. Consider what might happen as described in Table 6.2.

While the table is contrived to illustrate a point, data like this were observed in a project by the author and a colleague. Looking at the first need area, we see that for our NAC of 20 people, 10 saw the need as highest priority and 10 viewed it as lowest, yielding an average rank of 3.0. The second area achieved the same average with quite a different pattern of responses underlying it.

Table 6.2   Example of Rank-Ordering Possibilities in Needs Assessment

Need Area	Ranks					Average Rank	Rank Order
	1	2	3	4	5		
1	10	–	–	–	10	3.0	1.5
2	4	4	4	4	4	3.0	1.5
n							

Note: The numbers in the cells represent the individuals ranking the need area from 1 to 5. The average ranks were determined by multiplying the respondents in a cell by the corresponding rank in the column heading. In this example, 1 represents the highest importance for a need.

Reporting averages can sometimes mask what might be unique perspectives coming from subgroups within a larger one. More important, it might lead to probing into reasons for their choosing certain ranks over others. If there are subgroups and differences emerge, they should be examined. For the first need, why did half the group see that area as absolutely critical and the other believe it should be lowest in priority? This is precisely why averages should be accompanied by the distribution of responses.

Results like this were actually observed in rankings on a survey. They were reported and highlighted in a summary made to a decision-making board as input to deliberations and led to a healthy and enlightening discussion of different viewpoints.

3b. Simple Rank Ordering With a Few Additions

The entire explanation of rank ordering was based on the premise that the list of needs is short—15 or fewer. This is the normal state of affairs, but there are exceptions that take us to three small yet meaningful variations on the theme of ranking—*additional rules, ranking twice with a twist,* and *going through the list of items twice and only ranking subsets.* Some of these strategies are best seen through the practical example given below (see Example 6.1).

Example 6.1

What Can Happen With Rank Ordering?

A statewide organization was looking at its needs and possible solutions to them as part of a prioritizing process. Predicated upon identified needs, many solutions or courses of action were considered, and a complete list was generated after much discussion. Key solutions were placed on a wall and displayed in terms of major categories or clusters with individual strategies shown below them.

From there, members of the group were assigned magic markers of assorted colors and told that they had 10 points that they could assign in whole number amounts in any way they chose to any of the strategies. They got up from their seats and began in a quiet (with very little discussion) and thoughtful manner to place colored dots beside the solutions (or priorities) they felt were best suited for resolving needs.

The process was efficient, requiring only about 15 min or less to complete. It was quite revealing about the way members of the group were thinking and their strategies for assigning colored dots to priorities. Some used a larger number of the 10 points for one or two key priorities whereas others spread theirs out a lot more. This was evident by looking at the array of colors. It led to a spirited discourse about what factors were influencing decisions.

It might also indicate that some direction be given so that a person cannot allocate all or most of his or her points to one choice. By changing the rules, it seems that more options for decision making would be illuminated and group members would be forced to think about additional alternatives for problem resolution.

Another subtle observation was that selection often closely followed earlier group interchanges where there almost was an under-the-surface positioning and advocating for certain solutions at the expense of others. Group members were to varying degrees locked into already taken stances about issues and priorities without considering and fully thinking about the entries on the entire list. While not true for all, preconceptions governed the responses of some group members.

The example offers the NAC and its facilitator a lot to think about with regard to prioritization. Certainly, limiting the points assigned to a choice (an *additional* rule) would require more introspection, and choices would have been beneficial. Putting votes on just one or two choices doesn't require as much intellectual effort. It doesn't take into account that valuable and creative work went into the process of generating alternative solution strategies. It may show laziness on the part of individuals; they are tired and just want to get through this step in the assessment procedure. Fine and good! But openness in thinking is to be cherished and promoted. It is more challenging to one's own ideas. It is better to do this now than to have to come back to it later. A way to move forward and to break insular thinking is to do the following:

- Have the group members rank-order as above.

- Then, based upon what each person has done (as determined through colored choices), have them reselect, but now do *not* allow them to include their top priority in the ranks.

- Compare the two sets of rankings.

This *ranking twice with a twist* is similar to what is done in some balloting or voting procedures. This approach produces a new election with first choices deleted. When they are not allowed, a novel and different set of priorities might emerge. If agreement does not occur the first time around but does when each committee member cannot pick his or her first choice, then it might be that there is an important need that the group could support without too much difficulty. Not a lot of time would be needed to do this second ranking, and it is offered for consideration in this context.

The last option for rank ordering applies when there are long lists of items or needs. In a study, two surveys, for current and future needs, were used. Each was moderately long with over 30 items apiece. A way to handle this was to have respondents go through each survey a first time noting the items that were most important and those that were least important. They were instructed to look at the sublists and then pick the top seven that stood out and the bottom seven. For the top seven, they were asked to place them in order from 1 (the most important) to 7 (the least important of the seven). The bottom items were just left as an identified set of weaker needs. This *going through the list of items twice and only ranking subsets* was done for the two surveys, respectively.

The analysis consisted of counting how many times an item was selected in the top seven, what ranks it received, and what was the distribution of responses for each item. A decision rule in play here was that an item had to be selected by at least a minimum number of respondents to be in consideration. The number of times an item fell into the bottom seven was also recorded to see if there was some consistency as to what areas were least important.

On the surface this seems like a complicated procedure and one that would be difficult for respondents. The respondents were managers in corporations ranging from very large to small in size. Possibly due to their sophistication, they were able to follow the directions and easily completed the task. Return rates were high, so with appropriate and clear direction as to how to go about the process, no snags were noted. With proper preparation, this approach to rank ordering worked well.

Step 4: More Complex Prioritization

In reality all of the ideas so far have been about basic causal and prioritization strategies. They will suffice for the majority of needs assessment endeavors. But as described for causal analysis and as hinted at in a subtle manner for prioritization, they are not always appropriate. Some needs are complex, and with data coming from multiple methods and levels of need, the information will have to be carefully integrated and collated before causes and priorities are considered.

Some more complicated prioritization processes have been dealt with more intensively in Book 1 of the KIT. Rather than reinvent the wheel, the reader is referred to that source, and what will be added here is a little more background about detailed prioritization.

The key to prioritizing when the needs are complicated lies in defining criteria to guide the process. This is not done as often or fully as it should be in some areas (perhaps in relation to educational and social programs). Usually all that is asked for is some sort of ranking against a vague concern of what seems to be most important.

Go back to the general directions for ranking that were specified earlier. A number of criteria were linked together in one set of directions or imperatives to guide the process. That will be insufficient for needs that are complex. The Sork (1998) procedure in the first book is based upon making multiple criteria clear and rating each need on each one of them. Sork had two clusters of criteria—importance (five items for rating) and feasibility (three items).

Other criteria could be added in such as risks in resolving or not resolving needs, motivation in organizations, and level of cooperation across agencies participating in a collaborative needs assessment. Even others would fit specialized situations. Consider a college in a university that received the go-ahead from the central administration for the hiring of new faculty. Factors to guide choices might include:

- maximizing areas of strength that would continue to enhance the reputation of the college and its ability to attract funds (this same kind of thinking could be done in a business when it decides to eliminate a product line or expand another one);

- investing in weak areas that still have viability but have been dropping in quality;

- putting resources primarily into programs that are expected to emerge in the near future but are not fully eminent;

- spreading resources out over a variety of initiatives, thus keeping more personnel happy by not appearing to favor one group over another (notion of minimizing political risk); or

- focusing on one area this year in accord with a longer-term development plan that would then phase in new positions the next year and so on.

Remember the needs in this setting are involved with many issues that should be openly discussed before coming to final decisions. The future of the college could rest upon the criteria used as the basis for what should be done. Furthermore, note the political issues inherent in the criteria just explicated.

This was an actual situation in which the author participated. The decision makers heard presentations from units requesting new positions—all of them were cogent and persuasive. There were far too many new faculty hires being sought that somehow had to be sorted through, and then hard and binding choices had to be made. They were made, but to a high degree it was done in the absence of a well-thought-out, delineated, and communicated strategy.

Perhaps a better method for doing this would have been to use the *ranking twice with a twist* strategy suggested for selection of an area or item in which the rankers after first prioritizing go back through the process but now are not allowed to select their initial choices. Then, if results from this process were communicated, it would have conveyed more openness and fairness in the endeavor rather than an arbitrary tone to it.

On the other hand, it could have led to choices that might not have been satisfactory to many factions in the college. (Of course, individuals could always game the system by not choosing their first choice in the first round of selection.)

The upshot of this discussion is that in prioritizing of complex needs, explicated and well-defined criteria are necessary. Think carefully about important factors that affect decisions and choices and incorporate them into the criteria. To facilitate the process of selection, it may be best to keep the number (four to five) of criteria small. Lastly, it is always good practice to communicate the basis upon which final choices were made.

❖ ODDS AND ENDS ABOUT PHASE II

What About the Three Levels of Need?

It seems that they are nearly forgotten in the entire text; they are not, but the preponderance of emphasis has been on Level 1. So what about Levels 2 and 3? There are some choices confronting the NAC and its facilitator. It goes without saying that as Level 1 needs are assessed and preeminent in the needs assessment process, a lot will be learned about Levels 2 and 3 and solution strategies. So think about the following:

- How much do we know about Level 2, the service providers, and their discrepancies and gaps for service provision?

- How much is known about the overall system (Level 3)?

- Do we have enough knowledge that has been directly assessed or has arisen in the course of assessing the needs of Level 1?

- If the answers to the above questions are mostly in the negative, does the NAC feel that it must do a separate needs assessment for one or both of the levels?

- If we have a good understanding of what is going on at these levels, would it be useful to collect a little more information about them by applying some techniques (a few focus groups, a small number of interviews, going back to the literature, etc.) that can be done for not much expense and rapidly?

- After doing this, what is our feeling about a full Phase II needs assessment for Levels 2 and 3?

- Have ideas about solution strategies been seen, what are they, and how do they relate to causal factors?

- To what extent does all of this information affect overall perceptions of the need area?

Obviously, the hope is that enough has already been gathered that will help in answering most of these questions without too much investment of time, money, and human resources. (Parenthetically, is it possible to overstudy a problem or gap?) So it seems best to review knowledge to date and find out how the NAC feels about its understanding of the area of focus. Much of the necessary information probably is there. But the scope of the assessment has a lot to do with whether or not more in-depth study will be required.

In *Needs Assessment: An Overview* (Book 1), an example of a national assessment was presented that was vast (*national* is to be emphasized) in size. The agency carrying out the assessment felt it was better to do it for Level 1 initially and move to Levels 2 and 3 subsequently. With many potential different service providers, the agency's modification of the process was reasonable. Conducting a needs assessment is itself a relative type of exercise.

Models of Needs Assessment

Although one model guides this text, there are others that have unique features as well as features that overlap with it. Some of this has been partially covered in Book 1. For other models and analyses of them, see the Web site developed by R. Watkins (http://home.gwu.edu/%7Erwatkins/articles/NAdigest2.pdf).

Cultural Sensitivity

Not much has been mentioned about culture and how it affects Phase II activities. Since the impact of culture on an assessment can be dramatic, cultural factors have to be taken into account. One way would be by conducting a cultural audit of the organization that is asking for an assessment. It helps the facilitator and the NAC to learn about how the organization operates, communication flows, openness to ideas, who is involved or excluded from endeavors, what type of information would resonate with staff and decision makers, "sacred cows," and other related issues. This information would be invaluable for a successful study of needs. For details of the audit, see Book 2 of the KIT.

Another way to gain understanding would be through the literature. White (2005) did exactly this in a study of minority student retention in science and mathematics. He found that eight cultural propositions (Kuh & Love, 2000) may influence how minorities see the university and whether they persist in their studies. This knowledge led to the development of more effective and appropriate instruments. Gilson, Depoy, and Cramer (2001) generated, from a review of articles and research, an assessment model for the abuse experiences of women with disabilities. Not surprisingly, the topic is a touchy one, and a prominent feature of their approach was the incorporation of a cultural sensitivity component.

A third suggestion would be to include individuals from a specific culture in the design of the needs assessment and the instruments used for data collection. Carlson et al. (2006) approached the needs of a minority group for health information related to diabetes in just this manner. They also used focus groups with a sample of the population to obtain subtle insights into the issue. They suggested that this led to a much better assessment of the needs of the population. Pertinent here is the example of a focus group conducted with only female international Asian students at a major university (Chapter 5) and the interesting and unexpected findings it produced.

It is unlikely that any assessment can be implemented free from the culture of organizations, groups, and individuals with needs. When starting the needs assessment, be alert to culture and how the assessment must be tailored to fit its contours and shapes. By this means, the quality of the process will be enhanced, and more meaningful information will be collected.

Data Analysis

How to analyze data has not been treated to a great extent in this text. Note that Book 4 is devoted to this topic. It looks at qualitative and quantitative data and the combination of the two types. For the details of analysis, the reader is referred to that source.

Dated Tables

Keep in mind that needs assessment is a process that could be visually characterized as a funnel. It starts out wide at the top and narrows as it progresses down the incline. This is true from initial inception through all Phase II activities down to the culmination in Phase III of organizational plans for solution strategies and the evaluation of all the work done in the assessment.

Dates should be placed on all tables, and summaries should be produced in the course of the work to document the process. Dating tends to demonstrate the funnel effect. This is good practice and provides a record through which the needs assessment can be communicated to decision-making groups and key stakeholders and should be prominent in all communications. As the work progresses, the tables become a fount of information for all aspects of the process. They are good for taking stock of what has been learned and accomplished.

Another reason to do this is that the dated materials become part of the permanent record for later needs assessors long after this study is completed. Electronic and hard-copy files should be maintained. Most likely, needs will have to be periodically reassessed, and the value of having this type of record available and accessible cannot be overestimated.

Highlights of the Chapter

1. Stress was placed on briefly examining what has been found out about causes from prior needs assessment work.

2. A suggestion was made to start with causal analysis first and then to prioritize, but the NAC may want to reverse the order of these steps.

3. The fishbone technique was described as a good way for the NAC to look at causes, but if the situation is complex, fault tree analysis or other techniques might be better.

4. After causes are specified, simple forms of prioritizing (ranking procedures) were described with some discussion of varied ways of implementing this activity.

5. If the need area is complex, deliberate about criteria for prioritizing and make sure that they are transparent and communicated to all involved parties.

6. The necessity of good record-keeping procedures was underscored in this Phase as in other phases. The phases are interconnected and build off of each other.

References

ABLE Design and Evaluation Project. (2008). *ABLE collaborative needs assessment.* Columbus: Center on Education and Training for Employment, The Ohio State University.

Adler, C. L., & Zarchin, Y. R. (2002). The "virtual focus group": Using the Internet to reach pregnant women on home bed rest. *JOGNN, 31*(4), 418–427.

Altschuld, J. W. (2003, Summer). *Workshop for Korean educators.* School of Educational Policy and Leadership, The Ohio State University–Columbus.

Altschuld, J. W. (2004). Fault tree analysis. In S. Mathison (Ed.), *Encyclopedia of evaluation* (p. 154). Thousand Oaks, CA: Sage.

Altschuld, J. W., Anderson, R., Cochrane, P., Frechtling, J., Frye, S., & Gansneder, B. (1997). *National evaluation of the Eisenhower National Clearinghouse for Mathematics and Science Education: Final technical report.* Report submitted to the clearinghouse, The Ohio State University–Columbus.

Altschuld, J. W., & Austin, J. T. (2006). *External evaluation of the federally funded grant to the National College Access Network.* Cleveland, OH: National College Access Network.

Altschuld, J. W., & Kumar, D. D. (2009). *The needs assessor's handbook.* Thousand Oaks, CA: Sage.

Altschuld, J. W., Ramanathan, H., Ou-Yang, Y., Barnard, B., & Holzapfel, M. L. (1996). *Evaluation of the I Know I Can program.* Report submitted to the I Know I Can Foundation, Columbus, OH.

Altschuld, J. W., & Thomas, P. T. (2009, in press). Shifting sands in a training evaluation context: An exercise in negotiating evaluation purpose and methodology. In J. Morrell (Ed.), *Unintended consequences case book.* New York: Guilford.

Altschuld, J. W., White, J. L., & Lee, Y.-F. (2006). *Summary of OSEA evaluation activities and results.* Report submitted to the Ohio Science and Engineering Alliance, The Ohio State University–Columbus.

Altschuld, J. W., & Witkin, B. R. (2000). *From needs assessment to action: Transforming needs into solution strategies.* Thousand Oaks, CA: Sage.

Altschuld, J. W., Yilmaz, H. B., Harpster, A., Pierson, B., & Austin, J. T. (2006). *Summary and analysis of the focus group interview process: Evaluation of public health grand rounds.* A report for the Ohio Center for Public Health Preparedness, The Ohio State University–Columbus.

Bhopal, R. S. (2002). *Concepts of epidemiology: An integrated introduction to the ideas, theories, principles, and methods of epidemiology.* Oxford, England: Oxford University Press.

Bugeja, A. L., & Hwang, S. W. (2000). Barriers to appropriate diabetes management among homeless persons in Toronto. *Canadian Medical Association Journal, 163*(2), 161–165.

Burton, L. J., & Bruening, J. E. (2003). Technology and method intersect in the online focus group. *QUEST, 55,* 315–327.

Carlson, B. A., Neal, D., Gayenell, M., Jenkins, C., Givens, M., & Hossler, C. L. (2006). A community-based participatory health information needs assessment to help eliminate diabetes information disparities. *Health Promotion Practice, 7*(3), 213S–222S.

Centers for Disease Control and Prevention, National Center for Health Statistics. (1999–2002a). *National Health and Nutrition Examination Survey data.* Hyattsville, MD: U.S. Department of Health and Human Services. Available from http://www.cdc.gov/nchs/about/major/nhanes/datalink.htm

Centers for Disease Control and Prevention, National Center for Health Statistics. (1999–2002b). *National Health and Nutrition Examination Survey questionnaire.* Hyattsville, MD: U.S. Department of Health and Human Services. Available from http://www.cdc.gov/nchs/about/major/nhanes/datalink.htm

Chauvin, S. W., & Anderson, A. C. (2003). *Training needs assessment questionnaire for public health competencies in emergency preparedness and bioterrorism.* New Orleans, LA: South Central Center for Public Health Preparedness, Tulane University.

Chauvin, S. W., Anderson, A. C., & Bowdish, B. E. (2001). Assessing the professional development needs of public health professionals. *Journal of Public Health Management Practice, 7*(4), 23–37.

Chiasera, J. M. (2005). *Examination of the determinants of overweight and diabetes mellitus in U.S. children.* Unpublished doctoral dissertation, The Ohio State University–Columbus.

Chiasera, J. M., Taylor, C. A., Wolf, K. N., & Altschuld, J. W. (2007a). Correlates of diabetes in U.S. children from the 1999–2002 National Health and Nutrition Examination Survey. *Clinical Chemistry, 53*(6), A199.

Chiasera, J. M., Taylor, C. A., Wolf, K. N., & Altschuld, J. W. (2007b). *Correlates of diabetes in U.S. children from the 1999–2002 National Health and Nutrition Examination Survey.* Poster presentation at the annual meeting of the American Association of Clinical Chemists, San Diego, CA.

Chiasera, J. M., Taylor, C. A., Wolf, K. N., & Altschuld, J. W. (2008). *Correlates of diabetes in U.S. children from the 1999–2002 National Health and Nutrition Examination Survey.* Unpublished manuscript.

Cohen, B. (2006). Population health as a framework for public health practice: A Canadian perspective. *American Journal of Public Health, 96*(9), 1574–1576.

Conklin, N. L. Hook, L. L., Kelbaugh, B. J., & Nieto, R. D. (2002). Examining a professional development system: A comprehensive needs assessment approach. *Journal of Extension, 40*(5), 1–9.

Crane, M. (2009, January 29). CDC adds Ohio to wide probe of salmonella cases. *The Columbus Dispatch,* pp. A1, A4.

Cullen, C., Denning, R., Haury, D., Herrera, T., Klapper, M., Lysaght, R., et al. (1997). *Case studies: Teachers' perspectives on reform and sources of information.* Technical report from the Eisenhower National Clearinghouse for Mathematics and Science Education, The Ohio State University–Columbus.

Dance, P., Brown, R., Bammer, G., & Sibthorpe, B. (2004). Aged care services for indigenous people in the Australian Capital Territory and surrounds: Analyzing needs and implementing change. *Australian and New Zealand Journal of Public Health, 28*(6), 579–583.

Davis, M. V. (2006). *Mississippi Department of Health after action report.* Chapel Hill: The North Carolina Institute for Public Health, The University of North Carolina.

Dawber, T. R., Meadors, G. F., & Moore, F. E., Jr. (1951). Epidemiological approaches to heart disease: The Framingham study. *American Journal of Public Health, 41*(3), 279–281.

Eastmond, J. N., Jr., Witkin, B. R., & Burnham, B. R. (1987, February). How to limit the scope of a needs assessment. In J. Buie (Ed.), *How to evaluate educational programs: A monthly guide to methods and ideas that work* (pp. 1–6). Alexandria, VA: Capitol Publications, Inc.

Fairclough, G. (2009, January 27). Beijing goes on alert after avian-flu death. *The Wall Street Journal,* p. A7.

Friedman, D. J., Cohen, B. B., Averbach, A. A., & Norton, J. M. (2000). Race/ethnicity and OMB directive 15: Implications for state public health practice. *American Journal of Public Health, 90*(11), 1715–1719.

Garibaldi, B., Conde-Martel, A., & O'Toole, T. P. (2005). Self-reported comorbidities, perceived needs, and sources for the usual care of older and younger homeless adults. *Journal of General Internal Medicine, 20*(8), 726–730.

Gilson, S. F., Depoy, E., & Cramer, E. P. (2001). Linking the assessment of self-reported functional capacity with abuse experiences of women with disabilities. *Violence Against Women, 7*(4), 418–431.

Goering, P., & Lin, E. (1996). Mental health levels of need and variations in service use in Ontario. In *Patterns of health care in Toronto to the ICES Practice Atlas* (2nd ed., pp. 1–21). North York, Ontario, Canada: Institute for Clinical Evaluative Sciences.

Greenbaum, T. L. (2000). Focus groups vs. online. *Advertising Age, 71*(7), 34–35.

Hamann, M. S. (1997). *The effects of instrument design and respondent characteristics on perceived needs.* Unpublished doctoral dissertation, The Ohio State University–Columbus.

Hebel, J. R., & McCarter, R. J. (2006). *Study guide to epidemiology and biostatistics.* Sunbury, MA: Jones and Bartlett Publishers.

Hites, L. (2006). *Needs assessment perspectives from Tulane: What happened after Hurricane Katrina?* Panel presentation at the annual conference of the American Evaluation Association, Portland, OR.

Holding, R. (2007). Forced into the spotlight. *Time, 169*(5), 52.

Holton, E. F., Bates, R. A., & Naquin, S. S. (2000). Large-scale performance-driven training needs assessment: A case study. *Public Personnel Management, 29*(2), 249–265.

Hung, H.-L., Altschuld, J. W., & Lee, Y.-F. (2008). Methodological and conceptual issues confronting a cross-country Delphi study of educational program evaluation. *Evaluation and Program Planning, 31*, 191–198.

Hunt, M. H., Meyers, J., Davies, O., Meyers, B., Rogers, K. G., & Neel, J. (2001). A comprehensive needs assessment to facilitate prevention of school drop-out and violence. *Psychology in the Schools, 39*(4), 399–416.

Ishikawa, K. (1983). *Guide to quality control.* Tokyo: Asian Productivity Organization.

Ismail, A. (2004). Diagnostic levels in dental public health planning. *Caries Research, 38*(3), 199–203.

Jones, M., & Limes, K. (1988). *Gone fishing.* Unpublished manuscript, The Ohio State University, Educational Services and Research Department, Columbus.

Kenny, A. J. (2005). Interaction in cyberspace: An online focus group. *Journal of Advanced Nursing, 49*(4), 414–422.

Knight, A., & Meek, F. (2003). Needs assessment: A tool for hospice prevention. *International Journal of Palliative Nursing, 9*(5), 195–201.

Kuh, G. D., & Love, P. G. (2000). A cultural perspective on student departure. In J. M. Braxton (Ed.), *Reworking the student departure puzzle* (pp. 195–212). Nashville, TN: Vanderbilt University Press.

Kumar, D. D., & Altschuld, J. W. (1999). Evaluation of an interactive media in science education. *Journal of Science Education and Technology, 8*(1), 55–65.

Kumar, D. D., & Altschuld, J. W. (2004). Science, technology, and society: A compelling context for United States-Canada collaboration. *American Behavioral Scientist, 47*(10), 1358–1367.

Lee, Y.-F. (2005). *Effects of multiple group involvement on identifying and interpreting perceived needs.* Unpublished doctoral dissertation, The Ohio State University–Columbus.

Lee, Y.-F., Altschuld, J. W., & White, J. L. (2007a). Effects of the participation of multiple stakeholders in identifying and interpreting perceived needs. *Evaluation and Program Planning, 30*(1), 1–9.

Lee, Y.-F., Altschuld, J. W., & White, J. L. (2007b). Problems in needs assessment data: Discrepancy analysis. *Evaluation and Program Planning, 30*(3), 258–266.

Marklund, L. (2004). *Paradise.* London: Simon & Schuster.

Moseley, J. L., & Heaney, M. J. (1994). Needs assessment across the disciplines. *Performance Improvement Quarterly, 7*, 60–79.

Ohio Department of Education. (2006). *2005–2006 school year report card.* Columbus, OH: Author.

Passmore, D. L. (1990). Epidemiological analysis as a method of identifying safety training needs. *Human Resource Development Quarterly, 1*(3), 277–291.

Schmidt, W. H., & Wang, H. A. (2002). "What role does TIMSS play in the evaluation of U.S. science education?" Chapter 2 in J. W. Altschuld & D. D. Kumar (Eds.), *Evaluation of science and technology education at the dawn of a new millennium* (pp. 23–48). New York: Kluwer Academic/Plenum Publishers.

Schneider, S. J., Kerwin, J., Frechtling, J., & Vivari, B. A. (2002). Characteristics of the discussion in online and face-to-face focus groups. *Social Science Computer Review, 20*(1), 31–42.

Sebastian, S. (2007, January 11). More students are correct grade level. *The Columbus Dispatch*, pp. C1–C2.

Sork, T. J. (1998, June). *Workshop materials: Needs assessment in adult education and training.* Workshop sponsored by the continuing education division of the University of Manitoba, Winnipeg, Manitoba, Canada.

Taylor, C. A., Wolf, K. N., & Chiasera, J. C. (2006). Correlates of overweight in U.S. children from NHANES 1999–2002. *Journal of the American Dietetic Association, 106*, 63.

Turney, L., & Pocknee, C. (2005). Virtual focus groups: New frontiers in research. *International Journal of Qualitative Methods, 4*(2), 1–10.

Tweed, D. L., & Ciarlo, J. A. (1992). Social indicator models for indirectly assessing mental health service needs. *Evaluation and Program Planning, 15*(2), 165–180.

University of North Carolina Center of Excellence for Training and Research Translation. (2006). *Training needs assessment summary report.* Chapel Hill, NC: Author.

White, J. L. (2005). *Persistence of interest in science, technology, engineering and mathematics: An analysis of persisting and non-persisting students.* Unpublished doctoral dissertation, The Ohio State University–Columbus.

Whitmore, S. K., Zaidi, I. F., & Dean, H. D. (2005, December). The integrated epidemiologic profile: Using multiple data sources in developing profiles to inform HIV prevention and care planning. *AIDS Education and Prevention, 17*(Suppl.), 3–16.

Wilkie, K., & Strouse, R. (2003). *Custom report prepared for OCLC Institute: OCLC library training and education market needs assessment study.* Burlingame, CA: Outsell, Inc.

Witkin, B. R. (1994). Needs assessment since 1981: The state of practice. *Evaluation Practice, 15*(1), 17–27.

Witkin, B. R., & Altschuld, J. W. (1995). *Planning and conducting needs assessments: A practical guide.* Thousand Oaks, CA: Sage.

Witkin, B. R., Richardson, J., Sherman, N., & Lehnen, P. (1979). *APEX: Needs assessment for secondary schools* (student survey). Hayward, CA: Office of the

Alameda County Superintendent of Schools. (APEX consists of survey booklets for students, teachers, and parents and an administrator's manual.)

Yilmaz, H. B., & Altschuld, J. W. (2008). *Use of two focus group interview formats to evaluate public health grand rounds: Methods and results.* Poster presentation at the annual meeting of the American Evaluation Association, Denver, CO.

Yoon, J. S., Altschuld, J. W., & Hughes, V. (1995, Spring). Needs of Asian foreign students: Focus group interviews. *Phi Beta Delta International Review, 5,* 1–14.

Index

Supporting researchers for more than 40 years

Research methods have always been at the core of SAGE's publishing program. Founder Sara Miller McCune published SAGE's first methods book, *Public Policy Evaluation*, in 1970. Soon after, she launched the *Quantitative Applications in the Social Sciences* series—affectionately known as the "little green books."

Always at the forefront of developing and supporting new approaches in methods, SAGE published early groundbreaking texts and journals in the fields of qualitative methods and evaluation.

Today, more than 40 years and two million little green books later, SAGE continues to push the boundaries with a growing list of more than 1,200 research methods books, journals, and reference works across the social, behavioral, and health sciences. Its imprints—Pine Forge Press, home of innovative textbooks in sociology, and Corwin, publisher of PreK–12 resources for teachers and administrators—broaden SAGE's range of offerings in methods. SAGE further extended its impact in 2008 when it acquired CQ Press and its best-selling and highly respected political science research methods list.

From qualitative, quantitative, and mixed methods to evaluation, SAGE is the essential resource for academics and practitioners looking for the latest methods by leading scholars.

For more information, visit **www.sagepub.com**.